Science, Religion, Politics, and Cards

Science, Religion, Politics, and Cards

Maurice James Blair

Synapsid Revelations Press Corporation

Science, Religion, Politics, and Cards
Copyright © 2023 Maurice James Blair
Hardcover ISBN: 979-8-9887470-0-0
Paperback ISBN: 979-8-9887470-5-5
Blair, Maurice James 1976-

Synapsid Revelations Press Corporation
9619 Meadowcroft Dr
Houston, TX 77063

(Printing outsourced.)

1. Science 2. Religion 3. Politics 4. Cartomancy
5. Technology 6. Effectiveness 7. Resiliency
8. History 9. Science--Nuclear Physics 10. Metaphysics
11. Memoir 12. Ethics I. Title

Preface

Here is a hybrid passage that copies, resequences, and edits portions of two June 1st, 2023 electronic mail messages that I sent to a consultant (whose name appears much later in this book):

To the best of my memory it was either somewhat late into the year 1994 or somewhat early in the year 1995 when I picked up a neuroscience book that happened to be sitting around in my vicinity on that occasion. That was for a course that I never attended, though someone else at the dormitory clearly had it there for a reason or multiple reasons. Although I have no conscious memory at this time of the title, author(s), editor(s). and publisher(s) of that textbook, I have a clear and distinct and absolutely certain memory of what I found in it that caused me such righteous indignation at the time. It made very clear statements to the impressionable college-age youth to which it mainly aimed itself alleging: a) that all religious and spiritual belief systems completely lack reality, b) that anyone who believes it possible to genuinely experience the supernatural is clearly mentally ill, c) that anyone who would dare to look into a mirror and not fully identify with the physical manifestation of himself or herself would clearly be mentally deficient, and d) that any reader who felt resentment and/or strong disagreement toward that book's assertions of the aforementioned should then take a cold shower and repeat to herself or himself over and over again that it was correct to make these assertions, until finding herself or himself agreeing with the book on each of these claims. Words cannot express how angry I felt at that time. I threw the book down extremely hard, then soon walked away from the situation. No one else was anywhere near me at the time, by the way; I was alone in an area of that dormitory in terms of the presence of observable human beings when encountering that book and becoming angry at a level off of all normally-conceivable charts. Many years later, I let go of much of the sense of righteous indignation and hostility toward that book at various instances of time on account of developing many theories about why they might have been justified to present such stuff to youthful people, yet I still find it that there are some core levels of rot in an American society that for whatever reasons found itself doing many of the things that it has done from the mid-1990s to today, and I consider the extremities of how that textbook presented brainwashing to impressionable youth as one of the smoking guns of that rot. Much of what I have worked on since then could be considered the research and development of tonics and other tools with which to provide opportunities to improve situations.

I continue to believe that if the United States of America can manifest effective and truthful use of The First Amendment to the U.S. Constitution, then it can have extraordinary opportunities to in the long run repeatedly course correct for whatever rotten and toxic elements might emerge from time to time, due to the capacity for tonics and tools in general to emerge with which to steer things and people and places toward the good, the better, and the best.

Transitional Commentary Between Parts I and II
Consider comparing and contrasting various possible definitions and alternative definitions for the word "**define**" and the word "**infinite**."

<u>Opening Recommended Definitions</u> for Purposes of Reading this Book (though not completely excluding the consideration and usage of alternative definitions)

cartomancy: 1. the study and usage of cards to better understand and coordinate reality, with or without any divinatory or fortune-telling elements, though often popularly associated with divination and fortune telling; this can also be with occult remote influence of events, without occult remote influence of events, or somewhere in between being with and being without such influence. 2. the coordination and/or communication of card spreads as part of serving awareness of actual and/or allegedly-possible relationships between phenomenal reality/realities and noumenal reality/realities.

soteriology: the entirety of all real or alleged means of religious deliverance and the study of that entirety.

<u>Part II</u>
Next, here is a copy of the aforementioned **First Amendment:**

"Congress shall make no law respecting an establishment of religion, or prohibiting the free exercise thereof; or abridging the freedom of speech, or of the press; or the right of the people peaceably to assemble, and to petition the Government for a redress of grievances."

(Source: https://www.archives.gov/founding-docs/bill-of-rights-transcript as accessed on June 3rd, 2023)

Part III

Now consider the following passage from Nicholas Bessaraboff and Claude Bragdon's 1920 English translation of P.D. Ouspensky's *Tertium Organum*:

"To master the fundamental principles of *higher logic* means to master the fundamentals of the understanding of *a space of higher dimensions*, or of the world of the wondrous."
(Second Edition in English, page 263)

Part IV. For another perspective, consider a centuries-old comment from Benjamin Franklin:

"This modesty in a sect is perhaps a singular instance in the history of mankind, every other sect supposing itself in possession of all truth, and that those who differ are so far in the wrong ; like a man traveling in foggy weather, those at some distance before him on the road he sees wrapped up in the fog, as well as those behind him, and also the people in the fields on each side, but near him all appears clear, tho' in truth he is as much in the fog as any of them."
(Note: Corroborated by multiple sources as attributable to that statesman and scientist; although the publisher of this work is declining to single out any one or more of those sources by name here, at least several should be easily identifiable and verifiable for most readers with reasonably open access to any of a large host of resources. Please bear in mind that although Franklin was reportedly referring to a sect sometimes called the Dunkers, there has been ample evidence of similar modesty of the degree of supposition of accessing absolute truth in diverse peoples elsewhere in history, though he was correct for the sake of emphasis and a reflection of the reality of

his limited contact with many parts of the world to use the phrase "perhaps a singular instance" to point toward the general extreme rarity of this with respect to sectarian sentiments historically and geographically typical of *Homo sapiens*. Some might wish to compare and contrast elements of his referenced description of that sect, including his North American contemporary Michael Welfare, and various expressions of ideas prevalent from some sectors of the Himalayan region of the world over the span of many centuries. Although that region of the world has had much sectarianism, it has also sometimes and in some ways exhibited profound nonsectarianism.

Also, in relationship with all of this, it might be fruitful to a given reader to consider how experiments in physics have demonstrated many repeated instances of measurements (of observable physical reality comparing inertial reference frames) closely or exactly conforming to the Einstein-Lorentz transformation, in which relative magnitudes affect the degrees to which common sense addition can range from being very true to being very untrue, coordinating some of the different observers of motions, locations, and time.

Part V. Back to the aforementioned Bessaraboff and Bragdon 1920 translation of Ouspensky's *Tertium Organum*:

'"In order to approach a clear understanding of the relations of the multi-dimensional world, we must free ourselves from all the "idols" of *our world*, as Bacon calls them, i.e., from all obstacles to *correct* receptivity and reasoning. Then we shall have taken the most important step toward an inner affinity with the world of the wondrous."'
(Second Edition in English, page 263)

Part VI. On a related note, please consider a June 28th, 2022 comment I made in response to the *Religia bez ściemy* page on Facebook having posted an article regarding mysteries and facts in and adjacent to The Ten Commandments on June 27th, 2022 (as the date referenced by U.S. Central Daylight Time per my access to that online platform):

"Readers can take it or leave it, yet Ouspensky in *Tertium Organum* at times related the holding on too tightly to any logic structure to being a form of the mind of the beholder putting an idol between him-or-herself and the ultimate."

Part VII
As yet another portion of this opening fanfare, consider a copy of an 11:23 AM U.S. CDT May 15th, 2023 post that I presented to the general public via Facebook:

'"In the early 1990s a TV station from Los Angeles, CA was part of a cable television lineup popular in El Paso, TX, and that station played the Twilight Zone episode "The Gift." I know because I watched it."'

....
....

Part XX
Final revisions of inner text prior to the commencement of the first printing: July 11-20, 2023.

....
....

Part VIII

Here is a list of a set of television episodes and movies that readers may wish to consider as optional prerequisites to proceeding with studying the subsequent pages of this book:

- *The Outer Limits* / "The Man Who Was Never Born" (1963)
- *Tales from the Darkside* / "Seasons of Belief" (1986)
- *The Twilight Zone* (1959-1964 Series) / "The Gift" (1962)
- *Flashforward* / "Blowback" (2010)
- *Chernobyl* / "Vichnaya Pamyat" (2019)
- *Never Forget Tibet* (2022)
- *2001: A Space Odyssey* (1968)
- *Citizen Kane* (1941)
- *Oblivion* (2013)
- *Once Upon a Time in the West* (1968)
- *High Plains Drifter* (1973)
- *Freaks* (1932)
- *Casablanca* (1942)
- *Masters of the Universe* (1987)
- *The Incredible Shrinking Man* (1957)
- *The Twilight Zone* (1985-1989 Series) / Season 1, Episode 15 (1986)
- *Invasion of the Body Snatchers* (1978)
- *Beneath the Planet of the Apes* (1970)
- *Enter the Dragon* (1973)
- *The Nature of Existence* (2009)
- *Invictus* (2009)
- *Salt* (2010)

Science, Religion, Politics, and Cards

TABLE OF CONTENTS

PRELIMINARY PROCEEDINGS

PART ONE: HISTORIES, CARTOMANCIES, BIOGRAPHIES, AUTOBIOGRAPHIES, POLICIES, BIOLOGIES, NUCLEAR PHYSICS, METAPHYSICS, AND PRESENCES

PART TWO: SELECT FOUR-DIMENSIONAL MODELS OF COORDINATING CONSCIOUSNESSES, PREMISES, ENERGIES, AND RATINGS

PART THREE: REGARDING THE AMPLIFICATIONS OF INITIATIVES AND RESISTANCES

....

APPENDICES AND INDEX

Introduction

Although some people at the time of reading this may resent the hell out of a specific religion or philosophy or some combination of religions and philosophies, or may feel total condescension or condemnation toward anyone openly avowing such belief(s), it does not necessarily mean that such people will be vindicated or anywhere near it in the long run.

You might strongly agree with that statement or you might strongly disagree with it, but, either way, if you have any strong reaction to it - whether favorable or unfavorable - then a careful study of the testimony and cartomancies of this work might serve you well in the long run.

Although there is much that I withhold from fully disclosing in this, there are many things in these pages that could afford diligent readers of any faith or no faith ample opportunities to become aware of psychological warfare and psychological peacefare skills that might otherwise take them several decades or several lifetimes to comprehend. That being said, rather than recommend for anyone to read it or portions of it as soon as possible, I recommend that potential readers trust their own intuitions, sensations, thoughts, and feelings to help with when the times, seasons, and purposes are lined up well enough for them to proceed.

Introduction II

Consider this: At the time of encountering this page, who, of any living human beings, would you choose if only one could be chosen to fight on behalf of the human race against one extraterrestrial alien in a psychological warfare duel?

In this hypothetical, someone of extreme superintelligence and superpower, perhaps some manifestation of divinity and/or some extraterrestrial puts you up to volunteering one person to go through this, with some high stakes involved. Would you choose a mixed martial arts competitor? Would you choose a high-level business person? Would you choose a super-high-level political or religious leader? Who would you choose?

Introduction III

There have been many times in the around-mid-2012-to-mid-2016 period and from around August 2021 onward that if I had faced such a scenario as described in Introduction II, then I would have volunteered myself to be the representative for the set of beings who observationally manifest as the species *Homo sapiens*. Actually, let it be known that I did indeed consider that scenario mentally many times from about mid-2012 onward, often volunteering myself as the answer.

Why even consider such an altercasting scenario, and why make such a bold choice? Much of this is rooted in how during portions of mid-to-late-2003 through April 2005 (and frequently since then), I repeatedly set out to build resiliency and secret superweaponry of interpersonal interactional capabilities. Oftentimes I would keep much of this under wraps, waiting for if and when the times would be right to unleash these kinds of things on people. Also, of course, waiting for times to be correct for unleashing these on any and all categories of beings.

Rather than focusing on making this into an autobiography, however, my publisher and I are focusing on making this into a collection of tools that could prove helpful to diverse audiences.

On another note, since this is a nonfiction work, I do solemnly swear to do my utmost with these pages to be absolutely honest with both myself and the readers about:
- what I remember with absolute certainty
- what I remember with high probability though not quite absolute certainty
- things about which I am significantly uncertain at the time of composing this work

- things about which I am significantly certain at the time of composing this work
- relevant models and theories of which I am aware, their limitations, and their capabilities
- giving a reasonable amount of credit to others where such credit is due

On who or what do I make such a vow to do my utmost, you might ask? THE ABSOLUTE REALITY, to whatever degree(s) exhibiting as YHVH, ADI-BUDDHA, TAO, SHUNYATA, GOD, G-D, THE ULTIMATE REALITY, THE SOURCE OF THE ULTIMATE REALITY, or any other expression of how anyone ever has with sincerity to some degree succeeded in calling on and/or otherwise connecting with THE TRUTH, THE ACTUAL, THE REAL, & cetera.

Introduction IV

This work consists of narratives and card spread impressions of underlying energies spanning the period from June 20th, Four Hundred Seventy-One Million B.C.E. to July 10th, 2023 C.E.

Sections that refrain from overt cartomancy express testimony and conjecture regarding issues, events, and factors relevant to those concerned about science, religion, public policies, interfaith relations, and the development of personal resiliency.

Finalization of most pages of this work: June 20th, 2023 to July 11th, 2023 (U.S. Central Daylight Time).

Introduction V

To all the naysayers of why anyone anywhere should read this book and why it should even exist, bear in mind that it features information in and adjacent to:

1) How some soldiers in the early 1950s ignored instruction that forewarned them of their participation in a live ammunition crawling exercise, and they immediately perished; whereas other soldiers who were there survived because they took heed of the forewarning. (That is as relayed by one of the eyewitness survivors to me long before first publication of this book. See page 96.) That demonstrates that ignorance can be lethal and paying attention can be life-saving.

2) How several things that I directly experienced over the span of multiple decades are among what many might consider to be outside approximately the middle 8,999,999-out-of-9-million range of human experience.

3) Why an outtake idea for the title of this book was "Paul the Apostle versus Paul Churchland versus The World."

4) Why I have sometimes been a swing voter and plan to be one again when I believe it appropriate.

A BRIEF RISK ASSESSMENT OF INDEPENDENT CREATIONS OF IDENTICAL INTELLECTUAL PROPERTIES OR NEARLY-IDENTICAL INTELLECTUAL PROPERTIES AND ADJACENT RISK AND REWARD ASSESSMENTS

....

The author of this work
and the publisher of this work
do hereby certify that portions
not stated as attributable to
any specific person and/or entity
(whether named or unnamed herein)
other than the publisher and the author...
are items that the author and the publisher
have created, researched, developed, and delivered
without any conscious awareness of copying them from
any other sources, except insofar as de minimus
amounts of memes pervading history are in some
portions ethically rehashed, recombined, and
reorganized as parts of this work.

PART ONE:

HISTORIES, CARTOMANCIES, BIOGRAPHIES, AUTOBIOGRAPHIES, POLICIES, BIOLOGIES, NUCLEAR PHYSICS, METAPHYSICS, AND PRESENCES

...

...

...

...

...

...

...

Chapter One:
Outline for a Possible Uranium-235 Tarot Deck

MAJOR ARCANA (consisting of 22 cards)

I. Unity   II. Duality
V. Religion   XX. Moment of Truth
            0. Nonduality &/or Multiplicity
            XXI. The Great Ultimate
III. Methodology   XVIII. Unknown
    IV. That of All Methods and No Methods
            XIX. Execution
VI. Passion   XVII. Excavation
VII. Pursuit   XVI. Law
VIII. Super Symmetry   XV. Super Chaos
IX. #   XIV. ~
X. Beginnings   XIII. Endings
XI. Modern Wisdom   XII. Ancient Wisdom

MINOR ARCANA [consisting of 56 cards]: Four Suits (Creation, Destruction, Preservation, and Commerce) with 14 cards each: Ace, 2, 3, 4, 5, 6, 7, 8, 9, 10, Servant, Manager, Master, and Conquerer

TRANSITIONAL ARCANA (consisting of 14 cards)

B. Beyond   N. North   U. Universal
W. West   C. Center   E. East
D. Dharma   S. South   H. Heartfelt
        Tetragram Models
                            Voidness Models
                !
                    ?
                        !!

Abbreviated formats sometimes used in this book to identify cards of that U-235 deck:

UNITY   DUALITY   RELIGION   XX. MMTR

NonDuality/MLTPLCTY   XXI. ThGrtUlt   III. MTHDLGY

XVIII. UNKN   PASSION   XVII. EXCVTN

IV. T.o.A.M.&N.M.   EXECUTION   AD

PURSUIT   LAW   ACtn   ACmc   AP   2D
VIII. SUPERSYM 2Ctn   2Cmc   2P   3D
XV. SUPERCHS   3Ctn   3Cmc   3P   4D
~   X.BGN   4Ctn   4Cmc   4P   XIII. END

                         NORTH
                         CENTER
                         EAST
                         SOUTH
                         WEST

XI. MDRN WSDM   XII. ANCT WSDM
5Ctn   6Ctn   7Ctn   8Ctn   9Ctn   10Ctn
             Srv Ctn   Mgr Ctn   Cnq Ctn
10D SrvD MgrD CnqD   5D 6D 7D 8D 9D

BEYOND   HEARTFELT   UNIVERSAL   DHARMA

?   TETRAGRAM MDLS   VOIDNESS MDLS

5P 6P 5Cmc   6Cmc   7P
Mgr Cmc   7Cm   MgrP   CnqP
8 Cmc   9 Cmc   10Cmc 10P 9P   8P
             Srv Cmc   SrvP
        MstD   MstCmc   MstCtn

                        !

                     Mst P

                     !!

Chapter Two:
SYMBOLIZING PERSPECTIVES FOR THE 471 MILLION B.C.E. TO 470 C.E. PERIOD USING URANIUM-235 TAROT SPREADS

Outline:

Spread(s) 471M470__

{filling the blank with capital letters as digits in
a base-26-numbering-system sequence, where
A corresponds with the digit for Zero,
O corresponds with a single digit
for the number 14-in-decimal,
I corresponds with a single digit
for the number 8-in-decimal,
and R corresponds with
a single digit for the number 17-in-decimal}:

- AA, AB, AC, AD, AE,
- AF, AG, AH, AI, AJ

Those correspond with what in decimal would generally
be understood as:
- 00, 01, 02, 03, 04
- 05, 06, 07, 08, 09

Transitional Note: "Commerce" can be understood to
happen with beings outside of what most might
officially think of as human economic activity, whether
between beings humans generally consider lower than
humans, beings humans generally consider higher than
humans, or beings they may consider unspecified, &c.

471M470AA

"June 20th, 471000000 B.C.E."

0. Nonduality &/or Plurality   Ace of Preservation

XXI. The Great Ultimate   V. Religion   2P

3P 4P 5P 6P 7P   4D 5D 6D

7Cmc   H. Heartfelt

Ace of Commerce

XVII. Excavation

471M470AB

"July 26th, 468000500 to September 27th, 468000499"

ACtn 2Ctn 3Ctn 4Ctn 5Ctn 6Ctn

AD   AP
2D   2P
3D   3P
4D   4P
5D   5P
6D   6P

6 Cmc

5 Cmc

10 of Creation

Tetragram Models

471M470AC

"Voidness Models"

Voidness Models

"A Hybrid of 4004004.$\overline{004}$ B.C., 4004.$\overline{004}$ B.C.,

4.$\overline{004}$ B.C., 0.$\overline{004}$ B.C., 0.000$\overline{004}$ B.C., et cetera"

XVIII. Unknown

471M470AE

"Tetragram Models"

Tetragram Models

VI. Pursuit   XVI. Law

VIII. Super Symmetry   XV. Super Chaos

XXI. The Great Ultimate

471M470AG

"Garlands and Gargoyles in the Year 305 Million B.C.E."

Ace of Creation

Ace of Commerce VOIDNESS MDLS Ace of Destruction

Ace of Preservation

....

Ace of Destruction

5P   VOIDNESS MDLS   5D

Ace of Creation

....

Conquerer of Commerce

....

TETRAGRAM MDLS

VOIDNESS MDLS

Ace of Creation

"252 Million to 249 Million B.C.E."

3 of Destruction
4 of Destruction
5 of Destruction
6 of Destruction
7 of Destruction
8 of Destruction
9 of Destruction
10 of Destruction
Servant of Destruction
Manager of Destruction
Master of Destruction
Conquerer of Destruction
Ace of Destruction

XV. Super Chaos
XVI. Law
XVIII. Unknown
XX. Moment of Truth
XXI. The Great Ultimate
Tetragram Models
Ace of Creation
Voidness Models
I. Unity
II. Duality

0. Nonduality &/or Plurality

"Artificial Intelligence, Natural Intelligence, Intellectual Intelligence, Emotional Intelligence, Supernatural Intelligence, Imaginative Intelligence, Mentalities, Physicalities, and Metaphysicalities in the 249 Million B.C.E. to 470 C.E. Period"

0. Nonduality &/or Plurality
I. Unity
Tetragram Models   Voidness Models
II. Duality
Conquerer of Commerce
Conquerer of Destruction
Conquerer of Preservation
Conquerer of Creation

Ace of Commerce
Ace of Preservation

2 of Destruction
2 of Creation

5 of Creation
5 of Destruction
5 of Preservation

8 of Destruction
8 of Preservation
8 of Commerce
8 of Creation

"Untitled"

0. Nonduality &/or Plurality

I. Unity

II. Duality

9 of Creation

9 of Preservation

9 of Destruction

10 of Creation

V. Religion

Chapter Three:
OUTLINE FOR A POSSIBLE PLUTONIUM-239 TAROT DECK

Largely similar to the Uranium-235 deck, yet with the following differences:

- The minor arcana suit of Revelation complements the suits of Creation, Preservation, and Destruction, rather than having the suit of Commerce complement those three other suits.

- Two of the major arcana pairings have their numeral correspondences reversed, resulting in the inclusion of:

XIV. #   IX. ~

IV. Execution
XIX. That of All Techniques and No Techniques
(also known as XIX. That of All Methods and No Methods)

- Changing from having 14 transitional arcana cards to having 16 transitional arcana cards, by augmenting the other deck's set of them through the addition of:

?!   I?

SYMBOLIZING PERSPECTIVES FOR THE 471 C.E. TO 1955 C.E. PERIOD USING SOME URANIUM-235 SPREADS, PLUTONIUM-239 SPREADS, AND HYBRIDS THEREOF

AA

Ace of Creation	10 of Creation
Manager of Preservation	9 of Creation
5 of Preservation	6 of Revelation
6 of Destruction	6 of Commerce
7 of Creation	2 of Commerce
8 of Commerce	10 of Revelation
3 of Destruction	
	!!

AB

?!   !   !!

One Complete Suit of Creation

          Two Complete Suits of Destruction

Three Complete Suits of Preservation

          One Complete Suit of Revelation

          Two Complete Suits of Commerce

!?   !   !!

          V. Religion   ?

          XVIII. Unknown

          XVII. Excavation

          ?

          ?!

          !?

          !

          !!

AC

One Complete ^{235}U Tarot Deck

Two Complete Extra Suits of Revelation

One Complete ^{239}Pu Tarot Deck

Five Extra Instances of !

Six Extra Instances of !!

AX

XX. Moment of Truth

XV. Super Chaos

Six Complete Suits of Destruction

XIII. Endings

Fifty-Six Additonal Complete Suits of Destruction

Seventy-Five Additional Instances of XIII. Endings

Six Hundred Additional Complete Suits of Destruction

Seventy-Nine Additional Instances of XIII. Endings

Six Additional Instances of XV. Super Chaos

9,000 Additional Complete Suits of Destruction

XX. Moment of Truth

XV. Super Chaos

Six Complete Suits of Destruction

XIII. Endings

V. Religion

Eighty-Eight Complete Suits of Creation

Ninety-Five Complete ^{239}Pu Tarot Decks

10,800 Additional Complete Suits of Revelation

XVIII. Unknown

X. Beginnings

Nine Complete ^{235}U Tarot Decks

A Preliminary Supplemental Ace of Revelation

Additonal Ace of Commerce

Nine Complete ^{239}Pu Tarot Decks

Additional Ace of Destruction

Additional Ace of Creation

Additional Ace of Preservation

A Second Supplemental Ace of Revelation

A Second Set of Nine Complete ^{239}Pu Tarot Decks

BA

One Complete Suit of Creation

Ace of Preservation

9 of Destruction
10 of Destruction
Servant of Destruction
Manager of Destruction
Master of Destruction
Conquerer of Destruction
Ace of Destruction

6 of Commerce
7 of Commerce
8 of Commerce
9 of Commerce
10 of Commerce

?!
!?
?
!
2 of Commerce
2 of Preservation
2 of Revelation
IX. #

BB

Twenty-Five Complete ^{239}Pu Tarot Decks

XX. Moment of Truth

XXI. The Great Ultimate

Five Complete ^{235}U Tarot Decks

VIII. Super Symmetry

IV. Execution

XIX. Execution

IX. #

XIV. #

BC

X. Beginnings

XVIII. Unknown

XI. Modern Wisdom
XII. Ancient Wisdom

XIII. Endings

XIV. ~

XV. Super Chaos

XVI. Law

XVII. Excavation

XVIII. Unknown

?

VI. Passion   Ace of Commerce

X. Beginnings   VIII. Super Symmetry

XII. Ancient Wisdom   XVIII. Unknown

5 of Revelation
4 of Revelation
3 of Revelation
2 of Revelation
Ace of Revelation
Conquerer of Revelation
Master of Revelation
Manager of Revelation
Servant of Revelation
10 of Revelation

2 of Commerce
3 of Commerce
4 of Commerce
5 of Commerce
6 of Commerce
7 of Commerce

8 of Creation
7 of Creation
6 of Creation
5 of Cretion
4 of Creation

3 of Preservation
2 of Preservation

Ace of Destruction

BE

10 of Preservation

9 of Preservation

8 of Preservation
8 of Commerce
8 of Revelation
8 of Creation

7 of Preservation

6 of Preservation
5 of Preservation
4 of Preservation
3 of Preservation
2 of Preervation

Ace of Presrvation
Servant of Preservation
Manager of Preservation
Master of Preservation

Ace of Commerce

Conquerer of Preservation

XII. Ancient Wisdom

Full Suit of Preservation

Full Suit of Commerce

VI. Passion

VII. Pursuit

III. Methodology

IV. That of All Methods and No Methods

X. Beginnings

IX. ~ IX. #

XIV. # XIV. ~

XI. Modern Wisdom

Three Full Decks of ^{239}Pu

Three Full Suits of Commerce

An Additional Instance of E. East
An Additional Instance of W. West
An Additional Instance of S. South
An Additional Instance of C. Center
An Additional Instance of N. North

An Additional Instance of D. Dharma

An Additional Instance of Conquerer of Revelation

CE

Five Complete ^{235}U Decks

Two Complete ^{239}Pu Decks

Twenty-Five Additional Aces of Revelation

One Additional Ace of Creation

One Additional Ace of Destruction

One Additional Ace of Preservation

?

!

!?

!!

CF

WEST	XI. MDRN WSDM	XII. ANCT WSDM	5Ctn	6Ctn	7Ctn

8Ctn		5D	BEYOND
9Ctn			HEARTFELT
10Ctn		AD	DHARMA
Srv Ctn			UNIVERSAL
Mgr Ctn		AP	6D
Cnq Ctn			7D

8D	9D	10D	SrvD	MgrD	CnqD

CG

?!

!!

10D   SrvD   MgrD   CnqD

Ace of Destruction

Master of Preservation

Conquerer of Commerce

Manager of Creation

Servant of Preservation

10 of Preservation

?

!

CH

!?

?

!

?!

!!

CU

XA

XIV. ~ 　　　　　　　　XIX. Execution

8 of Revelation

5 of Commerce 　　　　　　2 of Preservation

II. Duality 　　　　　　　　5. Religion

XXI. The Great Ultimate 　　XIII. Endings

XB

7,000 Instances of Voidness Models

2 Instances of Tetragram Models

III. Methodology

XIX. Execution

IV. That of All Methods and No Methods

X. Beginnings
XIII. Endings
Conquerer of Commerce

XXI. The Great Ultimate

6 of Commerce
7 of Commerce
8 of Commerce
9 of Preservation
10 of Destruction

Servant of Preservation
Master of Destruction
Manager of Creation

V. Religion
VI. Passion
VII. Pursuit

Ace of Revelation

XIII. Endings
V. Religion
XXI. The Great Ultimate
6 of Commerce
7 of Commerce
8 of Commerce
9 of Preservation
10 of Destruction
Servant of Destruction
Manager of Creation
Conquerer of Commerce
Ace of Revelation
X. Beginnings

XI. Modern Wisdom   XII. Ancient Wisdom

B. Beyond   N. North   U. Universal

W. West   C. Center   E. East

D. Dharma   S. South   H. Heartfelt

I. Unity

II. Duality

O. Nonduality &/or Multiplicity

XL

IX. #   XIV. ~

                                              !!

I. Unity   II. Duality
V. Religion   XX. Moment of Truth
          0. Nonduality &/or Multiplicity

                              XXI. The Great Ultimate
III. Methodology   XVIII. Unknown
IV. Execution
          XIX. That of All Methods and No Methods
VI. Passion   XVII. Excavation
VII. Pursuit   XVI. Law
VIII. Super Symmetry XV. Super Chaos

IX. ~   XIV. #
X. Beginnings   XIII. Endings
XI. Modern Wisdom   XII. Ancient Wisdom

B. Beyond   N. North   U. Universal

W. West   C. Center   E. East

D. Dharma   S. South   H. Heartfelt

Tetragram Models

Voidness Models

                    !
?   !?   ?!
                   !!

{40}

303 Complete ^{239}Pu Tarot Decks

5 of Commerce

3 of Commerce

Ace of Commerce

7 of Commerce

9 of Commerce

An Extra Ace of Creation

An Extra 2 of Creation

An Extra 3 of Creation

An Extra 4 of Creation

An Extra 5 of Creation

?

?!

!!

XX

20 Complete ^{235}U Decks & 21 Complete ^{239}Pu Decks

18 Extra Complete Suits of Commerce

19 Extra Instances of VI. Passion

25 Extra Instances of XIII. Endings

25 Extra Aces of Creation

25 Extra Instances of X. Beginnings

CHAPTER FIVE: A DEUTERIUM TAROT DECK POSSIBILITY

XY		XX
XZ	DDD	EEE
YX	CCC	GGG
ZX	MMM	HHH
AZ	BBB	YYY
ZA	MGM	GMG
DD	FGM	MGF
EE	ZYX	ABC
	LLL	
	WWW	

AN ALTERNATE ARRANGEMENT OF A
DEUTERIUM TAROT DECK POSSIBILITY

ZA	MGM	GMG
DD	FGM	MGF
ABC	ZYX	EE
ZX	MMM	HHH
YYY	BBB	AZ
	XZ	
	DDD	
	EEE	
	LLL	
	WWW	
XX		XY
YX	CCC	GGG

CHAPTER SIX: OUTLINE FOR A POSSIBLE TRITIUM TAROT DECK

STP	PTS	PTSD
SSRR	RRSS	DSTP
ADAM	MADA	VEV
EVE	XYZYX	CBA
BC	JS	AD
DA	SJ	CBZ
YA	WLW	LWL
STD	DTS	TTT

AN ALTERNATE ARRANGEMENT FOR AN OUTLINE FOR A POSSIBLE TRITIUM TAROT DECK

TTT	DTS	STD
LWL	WLW	YA
CBZ	SJ	DA
AD	JS	BC
CBA	XYZYX	EVE
VEV	MADA	ADAM
DSTP	RRSS	SSRR
PTSD	PTS	STP

CHAPTER SEVEN: SYMBOLIZING PERSPECTIVES FOR THE 1956 C.E. TO 1999 C.E. PERIOD USING A COMBINATION OF SPREADS FROM THE AFOREMENTIONED FOUR AND THEIR HYBRIDS

"YY, YVONNE, STYX, WANG, WENG, XU, & OLAJUWON"

0. Nonduality &/or Plurality

XVIII. Unknown

VII. Pursuit

VI. Passion

VIII. Super Symmetry

GGG

VEV

MADA

XYZYX

"YZ, zyxwvuts, hgfedcba, abcdefgh, stuvwxyz, and ZY"

I. Unity

Ace of Creation

Two of Creation

ADAM

EVE

Three of Creation
Four of Creation
Five of Creation
Six of Destruction
Six of Revelation
Five of Preservation

X. Beginnings

XV. Super Chaos

ZA

MGM

GMG

DD

FGM

MGF

"Foreman, Foremen, Backman, Backmen, and Middlemen"

I. Unity	?	?!
II. Duality	!?	X. Beginnings
Ace of Commerce		3 of Preservation
3 of Creation	!!	BBB
4 of Commerce	GGG	XY
5 of Commerce	III. Methodology	
7 of Commerce		XX
10 of Commerce		XV. Super Chaos
Ace of Creation	VIII. Super Symmetry	
XII. Ancient Wisdom	0. Nonduality &/or Plurality	
		XIII. Endings
		XVI. Law

"82301 as conceived ahead of time in the minds of select Gentiles, Jews, Protestant Christians, Roman Catholics, Hindus, Kashmiri Shaivists, Buddhists, Sikhs, Jainists, Deists, Zoroastrians, Agnostics, and Muslims"

500 Aces of Creation   500 Aces of Revelation

500 Aces of Destruction   500 Aces of Commerce

500 Aces of Preservation   X. Beginnings

XV. Super Chaos   VIII. Super Symmetry

XY   XX   BC   AD   YYY   WWW   TTT

XI. Modern Wisdom   XII. Ancient Wisdom

B. Beyond   N. North   U. Universal

W. West   C. Center   E. East

D. Dharma   S. South   H. Heartfelt

I. Unity

II. Duality

0. Nonduality &/or Multiplicity

"ZC in the Light of 480 B.C.E. and 480 C.E."

XY

20 Complete ^{235}U Decks & 21 Complete ^{239}Pu Decks

18 Extra Complete Suits of Commerce

19 Extra Instances of VI. Passion

25 Additional Complete Suits of Revelation

25 Extra Instances of VI. Passion

25 Additonal Instances of Ace of Revelation

26 Supplemental Instances of XII. Ancient Wisdom

3 Supplemental Instances of XI. Modern Wisdom

Five Additional Instances of XIII. Endings

Another Complete ^{235}U Deck

!?

!

!!

"Forewoman, Forewomen, Backwoman, Backwomen, and Middlewomen"

10 of Commerce   XV. Super Chaos

Ace of Creation   VIII. Super Symmetry

XII. Ancient Wisdom   0. Nonduality &/or Plurality

TETRAGRAM MDLS   VOIDNESS MDLS

5P   6P   5Cmc   6Cmc   7P
       Mgr Cmc   7Cm   MgrP   CnqP
8 Cmc   9 Cmc   10Cmc   10P   9P   8P
    Srv Cmc   SrvP
MstD   MstCmc   MstCtn

!

Mst P

!!

Ace of Creation	Ace of Preservation
Ace of Destruction	Ace of Revelation
Two of Destruction	Two of Revelation
Three of Destruction	Three of Revelation
Four of Destruction	Four of Revelation
Five of Destruction	Five of Revelation
Six of Destruction	Six of Revelation
Seven of Commerce	Seven of Preservation
Eight of Commerce	Nine of Preservation
Manager of Destruction	Manager of Revelation
Conquerer of Destruction	Conquerer of Revelation

BBB XX XY GGG

IX. #
IX. ~
X. Beginnings
XI. Modern Wisdom
XII. Ancient Wisdom
XIII. Endings
XIV. ~
XIV. #

"ZG and The Fires of 1666 C.E."

STP	ABC	CBA	PTS
XIII. Endings		Ace of Destruction	
XII. Ancient Wisdom		2 of Destruction	
3 of Destruction		4 of Destruction	
6 of Destruction		5 of Creation	
2 of Preservation		2 of Commerce	
PTSD		DSTP	
	II. Duality		XIV. ~

"ZH and Volcanic Activities in the Entire 471 Million
B.C.E. to 1999 C.E. Period"

II. Duality   XVI. Law   XIV. ~
III. Methodology   IX. ~

I. Unity
VIII. Super Symmetry   XV. Super Chaos
0. Nonduality &/or Plurality

XXI. The Great Ultimate

IX. #
XIV. #

XIII. Endings

X. Beginnings

"ZIPPERS AND FAUCETS, 1742-1999, as viewed from
5,241 Aces of Creation"

I. Unity
Ace of Creation
Two of Creation
ADAM
EVE
Three of Creation
Four of Creation
Five of Creation
Six of Destruction
Six of Revelation
Five of Preservation

Seven of Destruction

Seven of Creation

Eight of Preservation

XXI. The Great Ultimate

205 Complete ^{235}U Decks & 212 Complete ^{239}Pu Decks

187 Extra Complete Suits of Commerce

199 Extra Instances of VI. Passion

20,500 Extra Complete Suits of Revelation

Five Supplemental Aces of Creation

108 Supplemental Complete Suits of Preservation

"Intertwined states of 59 & 45 Revisited"

Six Hundred Aces of Destruciton
MGM
GMG
FGM
Six Hundred Additonal Aces of Destruction
MGM
GMG
FGM
MGF
Eight Hundred Aces of Preservation
GMG
MGF
10 of Commerce
10 of Revelation
Servant of Revelation
Conquerer of Revelation
10 of Preservation
10 of Creation
Nine Hundred Aces of Creation

ZW

2 of Destruction

MGM   GMG   FGM

GMG   MGF   GMG

EVE   ADAM

WLW   LWL   LWL   WLW

STD   DTS   DTS   STD

STP   PTS   STP   PTS   STP   PTSD

PTS   PTS   PTS   PTS   PTS   PTSD

MGM   MGM   MGM   PTS   PTSD

FGM   FGM   FGM   FGM   PTS   PTSD

WWW   WWW   WWW

DSTP   DSTP   DSTP   DSTP   DSTP   DSTP   DSTP

WWW   LLL   WWW   WLW   WWW

Ace of Preservation

Ace of Destruction

57 Complete Decks of Preservation

{4 differentiated copies of the same suit can be a complete deck}

ZX

27 Complete ^{239}Pu Decks

547 Complete ^{235}U Decks

I. Unity

II. Duality

100 Billion Aces of Destruction

!!

2,000 Complete ^{2}H Decks

540 Complete ^{3}H Decks

!

?

!?

Twenty Trillion Aces of Destruction

!

Ninety Trillion Complete Decks of Destruction

!!

ZY

245 Servants of Commerce
531 Servants of Destruction
440 Servants of Preservaton
432 Servants of Creation
442 Servants of Revelation

Ace of Preservation
Conquerer of Preservation
Master of Preservation
Manager of Preservaton

10 of Preservation
9 of Preservation
3 of Preservation
2 of Preservation

IV. That of All Methods and No Methods

XIX. Execution

INTERMISSION ONE

{Chapter 8}

An Outline for Intermission One:

0. Miscellaneous Excerpts That May Provide the Reader Some Bearings and Levity Prior to the Recounting of More Heart-Wrenching Dramatics

I. A Biography of One of the Author's Former Neighbors, Specifically, A Man Who Reportedly Killed Himself

II. A Biography of One of the Author's Family Members, Specifically, A Man Who Reportedly Died While Asleep Soon After a Hospice Worker Gently Comforted Him While He Was In a Debilitated State, A Little While Prior to When the Hospice Worker Called the Man Who Would Go On to Become the Author of This Book to Inform Him about What Had Happened

III. An Autobiography of the Author of This Book, Deliberately Omitting Some Details That Were Already Revealed in Parts of His Past Books

IV. Relating the Four Previous Portions of Intermission One to Select Questions, Answers, and Additional Questions About Portions of Reality

V. Additional Focus on Some Issues Regarding The United States of America, Its Relationships, and Its Mysteries

Section Zero of Intermission One

Somewhat late at night on June 20th, 2023, I called an acquaintance who has often been a friend and occasionally, when we are at odds with each other, a mere acquaintance.

By normal evaluations of the biological development of the genetic-entity physical presences of human bioorganisms, he is in a regular sense something like twenty-five or thirty-five years older than me.

Of course, older does not always mean wiser. Here I shall identify him as "An Elderly Liberal Male Who Drove a 1992 Toyota Corrola Frequently in the 2016-2022 Period and Shared Many Conversations With M.J.B." or "E92P" as an abbreviation of that person's alias.

By the way, I have frequently driven a 2014.5 Toyota Camry in recent years. The "2014.5" in the previous sentence is not a typo, by the way, as although some refer to it as a "2014," it is technically of a design between the setups of the normal 2014 Model Year and the setups of the 2015 Model Year.

As a quick aside, there was a time approximately in the year 2011 that he and I were exercising in a gym, and we shared an interesting conversation.

Here is a hybrid of a best attempt to remember portions of that circa 2010-2012 conversation and portions of adjacent conversations:

M.J.B.: With how you expressed that you are a liberal and how I mentioned that I am an independent moderate, this reminds me of something that happened in a conversation a few years ago with a woman I dated.

E92P: How did that go?

M.J.B.: There was this woman who would go by the name Steph, short for Stephanie, and, well, the main thing she seemed interested in with the dating was to try to convert me to believing in Christianity as the only true religion. So some people, and she may be among them, might consider that it was not really dating at all. Yet, here is something more interesting than whether or not a given beholder considers what happened to have been dating.

In mentioning to her how Buddhism is generally what I believe in as most closely corresponding with reality, as far as human-labeled systems go, the dynamics eventually led to how I brought up the climate change issue.

I said to her at some point something like, "What do you think about how scientists say that if the human race fails to manage stuff like carbon emissions well, then we could go extinct much earlier than we otherwise would?"

She said in response something very much like, "We don't have to worry about that. God is in control, and what we humans do to the environment is not going to determine if and when the human race goes extinct, because only God can control that kind of thing. What is important is for people to believe in Jesus Christ."

E92P (chuckling): Did you laugh out loud when you heard her say that?

M.J.B.: No, I addressed it carefully and methodically. I mentioned stuff along the lines of how if there is a level of reality that corresponds with The Christian God being in control of all sorts of this kind of stuff - or even everything - then even in that type of reality maybe the way we as human beings treat the environment might affect God's judgment of what to do to the entire human race... and when to unleash what. Also, I brought up the idea of if there could be a setup in which reality includes God arranging to ty together actions and consequences such that a bunch of how humans affect global warming, global cooling, and other climate processes could automatically, through natural sciences, affect possibilities of human extinction versus human survival. Still, she stuck with the idea of God controlling everything in a way that makes concerns for climate change totally or almost totally irrelevant to how people should conduct their lives.

E92P: I probably would have laughed at her really hard!

....

Something that some readers might find rather ironic about the aforementioned conveyance of the essence of some conversational snippets from long ago is that the referenced conversational partner has told me repeatedly that, in addition to being politically liberal, he was raised to believe in Methodist Christianity and continues to be a Methodist Christian.

....

Next, to a best attempt to accurately remember and relay to the readers a semblance of a few interesting portions of the related June 20th, 2023 phone call, a conversation in which the process of developing this book became one of the subjects of conversation:

E92P: Are you writing that book mainly for profit or as a hobby?

M.J.B.: Neither. The main aim is at communicating truth to people in an honest way such that it can push back against many of the problems in this country and problems with the human race in general.

E92P: If you're not trying to make money from it, then what are you after?

M.J.B.: Profit is one of the motives. Actually, this new book is being published by the C Corporation that I own, and since I can in some cases act as a customer of it to send information to people, that corporation could be guaranteed to make a profit at least some of the time. This ties in to what I told you before about the book I published in early January. Or that could be in another sense the four books I published in early January this year, because that book is in four formats: one paperback and three hardcover formats. Eventually, one of the things I did with that was to buy one gift copy each of it to arrange to go to two senators and two congressional committees.

Some people say to others messages in the form of, "We have Cause X, and our organization believes that you, too, should believe in Cause X. Therefore, we will try to convince you to agree with us about this, and we want you to contact the U.S. Congress about this issue. You really should agree with us, and we'll tell you why, and you should seriously consider contacting congress on behalf of this issue." Well, what I did by sending those copies of that book to portions of the U.S. Congress was to write to congress about all issues, all people, and all concerns.

E92P: If I was a congressman, I might set such a gift aside and barely get to it, since I would have so many other things to do.

M.J.B.: That could be possible and fine for some of them. However, even within the first 15 pages of that book, available for online preview at the website for one of the retailers, it is blatantly obvious to almost anyone that I am making intent commentary on many vital issues. In the time since you became aware, early this year, that the preview is available, have you even bothered to look at that preview?

E92P: No, I haven't.

M.J.B.: See, then that's part of why you don't know about that.

E92P: What do you think is the biggest problem in America at this time?

M.J.B.: I am not so sure that there is even one specific biggest problem at this time; it could be that there are several coequal problems at the highest level of difficulty.
　Also, it is probably deeper than anything obvious. I do not know what the biggest problem in America is at this time, or if there are several coequal biggest problems. However, I do know many of the intermediate level problems to a major degree, and, as many people are aware of, pushing back against problems can help uncover more hidden problems.

E92P: If you had to name just one thing that you know of as being a problem in this country as a pressing issue, then what would it be?

M.J.B.: The gap between the reality of scarcity and people's perception of not having that scarcity.

...

Part of what went on for a long time in the conversation after I brought up scarcity was that I clarified about tradeoffs, in which many people want what they consider to be better for themselves, their families, the country, and whoever and/or whatever toward whom/which they feel favorable while not giving a damn about how if reality were to actually be able to grant them much more of all of those things that they want, then reality would have to steamroll and rip asunder much of the well-being of other individuals, other families, other entities, and so on and so forth.

One example that I gave him, and which I believe appropriate and legal to include here in this work, is that I had seen at a local vehicle repair shop a sign with which the business makes a perfectly clear statement to its customers: Something almost identical to, "1) Fast Service, 2) High Quality Service, 3) Low Price. You can choose two out of the three, but you cannot choose all three."

At first, he indicated that he did not get it. I then proceeded to explain to him that if a customer is strongly intent on imposing on a car repair place that the business give them very fast, high-quality service at a very low price, while the business has its constraints of how many employees, what to pay those employees, how many other customers, time available to devote to the entire set of customers, tools, regulations, and other such factors, then that strong imposition is a real problem for the repair shop. What would it take for the shop to thoroughly satisfy and please a customer strongly intent on imposing all three on the shop? For almost any - or maybe every - repair shop, to satisfy a customer locked in to that mindset would be to steamroll other customers, its employees, its management, and/or its owners. If the company were to try too hard to satisfy customers with that kind of three-pronged intense drive of desires, then the company would be virtually obligated to drive its financial profitability into the ground and likely soon find itself in the scrap heap of destroyed companies.

(I remember that auto repair shop to have been in the International District of Houston {also sometimes referred to as the Chinatown area in part of Southwest Houston, though it includes a major presence of Vietnamese, Korean, and other Asian communities and cultures, including hybrids of multiple manifestations of Asiatic presence(s)}, though I do not remember with certainty at the time of composing this which of those shops it was. On a more important note, I believe that to whatever degree that the auto repair shop referenced or one of its vendors might have an intellectual property claim to the contents of that sign, my publisher and I can legally include the description here of that sign as part of the fair use of copyrighted materials; and to whatever degree that the message on the sign and/or some semblance of it are in the public domain, then my publisher and I are obviously acting with full compliance with intellectual property laws in the process of including it.)

The aforementioned male conversational partner and my mother (to whom many people refer via the name Ming Blair) often agree with each other on the side against there being much worth in my creating and offering books to the public.

Some of my recent projects are *The Dimetrodons, the Dorians, and the Modern World* (2022), *The Dimetrodons, the Dorians, and the Modern World: Revised Edition* (2022), and *All Things under and over the Sun and Stars: Enigmas in Various Stages* (2023).

I gave the E92P person a complimentary copy of *The Dimetrodons, the Dorians, and the Modern World*, and to the best of my memory the date of giving that to him was October 29th, 2022. In response to how he has since read some portions of it and complained to me about some of its radically unorthodox literary features (specifically, that he found the first chapter to be difficult reading and found it awkward that I placed a nonfiction section between the second and third chapters of the story narrative), I informed him that much of the design involved intentionally upending literary conventions. Much of that unconventionality was in order to do my utmost to be true to honoring the mysteries of the ancients.

....

Sometimes in recent months, I figured out that a way to save some energy and time is that when a call comes in at the landline at my residence and looks according to caller ID to likely be a telemarketer or a scammer, is to handle it in the following manner:

First, immediately after picking up the phone, intentionally hesitating for an unusually long time, though not way too long, specifically, a few extra seconds compared to normal answering.

Second, to start speaking; for example, with "Hello."

Third, to listen for whether there is a response.

(In this scenario in those recent months, there is almost never a response at that point.)

If there is a response, then deal with it on an ad hoc basis.

If there is no response within the normal range of time, then to simply state something identical to or very similar to, "The intentional delay on this end evidently helped speed dial or whatever else to not bother to pick up at the other end." Right after stating that, hang up the phone.

....

The Dimetrodons, the Dorians, and the Modern World is a novel which had an ISBN-registration-system publication date declared in advance to be planned for October 6th, 2022 before winding up with an actual publication date of September 29th, 2022.

The Dimetrodons, the Dorians, and the Modern World: Revised Edition is a novel which had its ISBN-registration-system and actual publication date matching with October 10th, 2022.

Both of those books refer briefly to how there had been a break from correspondence between Dorsey Armstrong and me from about the 2nd quarter of 1995 until March 22nd, 2010; however, this was only a half truth in many respects.

In December 2009 I sent an e-mail message to her while expressing an edge of uncertainty about whether she had been the University Writing Course instructor who instructed a class of students, including me, during portions of August to December 1994; in some sense that 2009 communication could be considered a Festivus Day 2009 act of writing without contacting (on account of the presence of significant uncertainty of identity), whereas in another sense it could be considered a Festivus Day 2009 act of writing with contacting (on account of a more regular way of interpreting basic reality).

...

One of the most valuable experiences in my life thus far occurred when my family took a trip from our then-residence of El Paso to go East quite a ways, with not as much Northward or Southward movement nearly as much as the Eastward movement, arrived at a small cavern tourist spot and took a tour. Although I did not memorize the name of the place at the time, I have later found through indirect evidence that Sonora, Texas is one likely candidate of where the place may have been.

I vividly remember seeing on the way into the main chambers a stuffed sloth or a very similar stuffed animal. One of the most memorable things from my entire life was how the tour guide turned the lights out, waited a little while, then turned the lights back on. When the lights were out, it became absolutely dark down there in that cave. Evidently, they managed to not have any remaining light sources during the temporary full exiting of light.

Although some philosophers have at times emphasized perspectives of fully identifying the self with the observable physical presence of a body, memories of experiences such as the Sonora-or-similar-cavern lights-out session curb excessive agreeability with the impositions of those philosophers. This is part of the reason that, with reasonable respect for a degree of primacy about consciousness over many ranges of reality, I at times strongly oppose dogmatic adherence by some portions of academia and some sectors of the medical establishment to reductive materialism, eliminative materialism, and similar types of metaphysical materialism. Of the multitudes of people and organizations that I have gone through in a rough and impactful manner like how Bo Jackson famously went through Brian Bosworth on his way to a rough touchdown long ago, each has in some ways when that happened stood in the way of my arriving at touching down with reasonably-thorough statements to score points against many of Paul Churchland's statements that had favored the aforementioned dogmatism.

Even though I do not remember the identity of the neuroscience textbook that I happened to serendipitously briefly encounter in a dorm (as described in the preface), I do remember that a huge part of why I felt such justifiable rage at its brainwashing was that I had studied portions of what Paul Churchland had expressed in support of either the same or a very similar dogmatism, when assigned by Philosophy Professor Tad Schmaltz to study those portions.

By June 3rd, 2005, with the help of attempting to channel the entirety of reality to in some ways anonymously and in some ways semianonymously deliver to the world a book which I would publish on the subsequent day, I had in some ways moved on, though in other ways it is only with this book that I am fully moving on from the repeated long-distance warfare with the Paul Montgomery Churchland reportedly born on October 21st, 1942 in Vancouver, British Columbia and having in some recent times professional business relationships with multiple educational institutions, including The University of California, San Diego.

Of course, in some respects this decades-old feud is only the tip of the iceberg of many much bigger and deeper feuds within the fabric of reality.

On the other hand, here are a few of other things: 1) On the day of the 2007 Indy 500, I successfully completed a degree of wholesale psychic-plane annihilation of eliminative materialism, reductive materialism, and similar materialisms as dogmas by turning many of their core energies upon themselves, via the short story, "1970 Public Access IV Nightmare" which I published for free access on Free Stories Center (freestoriescenter.com).

On the next few pages, there is a copy of that story, reformatted to show here clear labeling as a "fiction section... within a nonfiction book" and with minor additional editing.

(Start of a Fiction Section of pages 76-80 as an exhibit within a Nonfiction Book)

1970 Public Access TV Nightmare

The characters of this story are used fictitiously. Similarities to actual persons living or dead are coincidental.

"1970 Public Access TV Nightmare"

by M.J. Blair

Tom Singleton and Shirley Stapleton enjoyed Beatles and John Lennon music during the 1960s, and now faced a personal spiritual crisis in December 1970. They had celebrated mysticism, believed in the powers of The Occult and Eastern religions, and felt wonder in how songs like "Strawberry Fields Forever," "I am the Walrus," and "Instant Karma" related to these. However, they recently bought the new LP *Plastic Ono Band* and felt confused by John Lennon's disavowal of considering Magic, Jesus, Buddha, Krishna, and others as authorities in his life.

Tom and Shirley roomed together in Chicago, Illinois, and were in their early twenties. They both had late December off from work in 1970 and spent much time meditating on how the universe had recently produced the Nixon administration, the Vietnam War, and *Plastic Ono Band*. Late at night on December 23rd, they watched local public access television, still looking for an answer.

First, they saw a ten-minute episode of a home-made soap opera. In part of the story, a man named Bob came back from Vietnam and suffered from post-traumatic stress. He told his girlfriend Charlotte that he believed Richard Nixon was put into the presidency by divine intervention and was at heart a spiritually pure human being. His girlfriend told him he had too many drugs in Vietnam.

Second, they saw a bearded man in his thirties, a short-haired woman in her late twenties, and a tall, clean-shaven man in his early twenties speaking about space and time and consciousness. During this program, the bearded man said, "All of the universe, all time past, present, and future are one, and all are ever-present and ever-changing. The beginning of the war is present, changing, and affecting us, the middle of the war is present, changing, and affecting us, and the end of the war is also already present, changing, and affecting us. All space and time, or as some call it, 'space-time,' has presence in our minds and everywhere, and we can have access to everything through our consciousness."

The short-haired woman responded, "Then, Charlie, how do you respond to some viewers who may be thinking, 'But I know that I am not the same person as my cousin, or I know that there are many different people and sentient beings... how can you say that everything is one when there are many different things and different beings?' What would you say to those viewers?"

The bearded man answered, "The reality is beyond the limit of our concepts. The way of organizing the concepts to demonstrate a level of unity of all consciousness and space-time provides our minds with a set of tools that are valuable. The way of organizing concepts around the distinction between different people, different periods, and different, unique things also provides us with a valuable set of tools. Our minds have to leap beyond each perspective to see the truth in alternate perspectives."

The clean-shaven man commented, "For many people, what you have just said probably went over their heads and left them feeling disappointed that you were not able to communicate to them at their level."

The bearded man responded, "That is unfortunate, but there is no way I could avoid it. I hope some people out there will benefit from these discussions, although others may have changed the channel when the speech became unclear to them. I guess sometimes you can't please everyone."

Next, Tom Singleton and Shirley Stapleton saw the public access channel sign off for the night. Rather than changing the channel or turning off the television, they held each other close and felt their concerns with the state of the world and spiritual confusion dissolving into the night. A little while later, however, Shirley was shocked to see something strange on the television.

"Look at the screen, Tom! What is that?" she said.

"What in the world?" he responded.

On the screen there appeared a creature that looked like a mixture of aardvark and human, speaking on the screen while sitting on a throne. It spoke in English.

"People of Earth, beware of the transmission you are about to receive, and beware of the cosmic midnight about to befall you."

Next, the screen turned black, and eerie synthesizer music started. A voice announced, "You believe many things, but much of what you believe is a lie. You are not human beings, and there are no things that you call minds. You are entirely physical phenomena, and, in fact, you have no minds."

After this, the television showed moving shadows and dark colors shifting shape on top of news footage of the time. Music emerged from the TV, music familiar to Shirley and Tom, but with a change: it was the music of "Revolution 9" sampled and altered with the repeated phrase "Number nine, Number nine, Number nine..." edited out and replaced with "You have no mind, You have no mind, You have no mind..."

They watched in horror as the images mixed with shadows changed from news events to footage of Soviet and Nazi torture sessions and horrible human experiments, as all the while the music with the recurrent phrase "You have no mind" continued.

Finally, as this video ended, the partially-aardvark-shaped humanoid again appeared on the screen. It said, "What you just witnessed was a propoganda video produced by The Interstellar Reductional Materialist Empire and designed to attack your world. They have attacked my world numerous times. Although they have a small presence in your world, beware of their future attacks."

2) Years later, I considered more thoroughly how from various Buddhist and Taoist perspectives all idea structures - no matter how absurd from a given viewpoint - have, at least in limited contexts, at least a little degree of reality. That led to how I, in meditation of considering reality from perspectives of Paul Churchland and people of a similar ilk and from various other perspectives, suddenly encountered freer awareness of possibilities adjacent to normal formulations of gravity. I soon communicated with *the acc-list* within Yahoo! Groups regarding possibilites of multiplying Clarke's 4th Law by multiple dimensions of the conventional idea of gravity: such that its normally presumed, calculated, and measured presence, attraction, and formulated magnitude could be considered as not necessarily always applying.

In otherwords, to look at both sides of its presence-versus-nonpresence, attraction-versus-repulsion, and standard-formulation-versus-alternative-formulation. I received feedback from that list that this was both a radical approach to the issue of gravity and something that other people had already explored in some of the more esoteric areas of theoretical physics. Here is a copy of that post I made to *the acc-list* within Yahoo Groups as displayed (presumably according to time as measured in Houston, TX, because of the settings of registration; attempting to exactly replicate characters and punctuation, though with different font settings):

On 9/26/10 12:26 PM, Maurice wrote:
"Clarke's Fourth Law in Relation to Gravity"

[Prefatory Note: I don't know the full extent to which things involving the following have already been adequately discussed in this forum; although I've read many past messages in the acc-list, there are many more that I haven't read. If you know of the most relevant past posts, please don't hesitate to point out the message numbers and dates for them.]

There's an article at http://map.gsfc.nasa.gov/universe/uni_accel.html about the Cosmological Constant and Dark Energy. It mentions:

"Surprisingly, the results of these observations indicate that the universal expansion is speeding up, or accelerating! While these results should be considered preliminary, they raise the possibility that the universe contains a bizarre form of matter or energy that is, in effect, graviationally repulsive."

In some ways this could tie in to Clarke's Fourth Law. "For every expert there is an equal and opposite expert." Specifically, we typically lock into the idea of gravity that attracts based on rather rigidly relating masses and distances. Taking that and going into something similar to the NASA statement though not quite the same, I'll say this: What if it's not always just "a bizarre form of matter or energy that is... gravitationally repulsive," but also sometimes in some circumstances "ordinary matter?"

[Additional Note: I just had three main physics classes in my undergrad degree, there was only a little chemistry coursework, and my master's degree did not include any natural science classes. Therefore, if some list members with much more thorough backgrounds in organized natural sciences think I sound kind of funny in my writing style here, it seems quite understandable to me.]

Part of the idea I have with this is that if we apply Clarke's 4th law to the very foundation of gravity, then whether matter is "bizarre" or "ordinary," it may seem possible that when relations meet some unknown criteria (or some criteria to be known in the future... or criteria already known by some mysterious sentient beings), then the gravitational effect could be different than under the inverse square law. We could apply the idea of "opposite" under multiple axes, too.

Opposite along an axis of sign: Gravity could repel rather than attract, as NASA suggested in the linked page above.

Opposite along an axis of magnitude: Gravity could have a greater or a lesser magnitude than the standard law's formula would indicate, an idea which a number of people have incorporated in some ways, though I don't know much of the specifics on this.

Opposite along an axis of presence: Gravity could be absent, an idea which science fiction writers have employed with such frequency that some view it as a cliche.

Now, each of the above may seem somewhat pedestrian in isolation, but I suspect that some interaction of all three in relation to intergalactic scales of matter, energy, and space-time could wind up allowing a model where the universe may not require nearly as much dark matter and dark energy as the recently prevailing notions have tended to lean on.

Anyone wish to share any comments?

- Maurice J. Blair

....

After I posted that message, William Wheaton responded with a mixture of humor and insight, including reference to how there are multiple Modified Newtonian Dynamics (abbreviated "MOND") theories that some people have purported as modern theoretical competitors to classical General Relativity.

On another note, though Yahoo! Groups closed down about a little less than one decade after that communication exchange between Bill Wheaton and me, I still occasionally reflect on all past interactions with other people and conclude every now and then that if I had to vote for one group of people to be the most capable of averting a global disaster, then it would be the collection of each person who was ever a member of the acc-list portion of Yahoo Groups.

....

3) Quite a while after that, I started to sense that maybe the presence of "1970 Public Access TV Nightmare" on the Internet (for anyone with reasonable web access to read) could be problematic in the long run if not for the provision of a reasonable tonic to it. Therefore, I set out to write and indeed wrote a follow-up story to serve the public as a tonic to the prior story.

I posted that to the same online repository for many authors: freestoriescenter.com.

On the next few pages is a copy of it:

(Beginning of a Fiction Section of Pages 85-88 as an exhibit within a Nonfiction Book)

Camps of Camps and Angles

The characters of this story are used fictitiously. The author wrote it in the morning of January 20th, 2013. Any similarities to actual events before, during, or after January 2013 are what they are.

"Camps of Camps and Angles"

by M.J. Blair

At a February camp, a college sophomore and a college freshman spoke about feeling mystified by recent comments from a village elder.

The freshman said, "What am I to make of all this? After listening to that, I feel like I'm just some silly dust floating along, hardly knowing anything about anything. How do you feel about it?"

The sophomore said, "Really? All that philosophy may not make perfect sense to me, but it kind of makes sense. Still, the elder's speech had to have been mysterious to most of us. Maybe all of us."

The first-year student followed up by taking things
further. "I'm a human being. You're a human being. The
village elder's a human being. Yet, I think back and I'm
not entirely sure anymore. I guess that up through high
school I just didn't find out much about this
metaphysics debate stuff. This kind of idealism, that
kind of idealism, this kind of materialism, that kind of
materialism, this dualism, that dualism, this
nondualism, that nondualism, here a realism, there a
realism, here a theory, there a theory, everywhere lots
of theories. I don't know how practical it all might ever
amount to being for anyone."

The second-year student smiled, "Sometimes a little
ambiguity and mystery in things like this could
be *very* practical. For example, think about if some
writer out there wrote a story in which he or she makes
no reference to whether any of the characters are
children, women, men, robots, aliens, etcetera. Then
people just take the story and debate whether this
character or that character is a man, woman, child,
alien, robot, or whatever. This could still be a good mind
exercise for them to sharpen their survival skills."

This drew a smirk from the freshman, who said, "Come
on. I think you're stretching to find something practical.
I mean some of those theories seem to be telling me
that people lack minds or that people lack bodies or
both. *Get real.* I have a mind. You have a mind. I have a
body. You have a body."

The sophomore's face suddenly adjusted into having a rather calm and neutral expression to it, and from that student's *being* came a series of age-old questions: "What makes someone a human being? Are all of us just automatically human as our birthright, or can we gain, lose, and regain our humanity? What is a mind? Does it have a location, or some mixture of multiple locations, or something else? How do minds relate to bodies? Does a mind possess a body, or a body possess a mind, or both? Or is it sometimes one way and sometimes another way? I think you have to admit these are interesting questions."

The freshman looked annoyed and argued back, "Interesting only if we let them be interesting to us. But if you let just *anything* seem interesting at any time, that could take away from being practical. Also, if you let things be interesting to you too easily, someone can come along and more easily finagle you. Less practical. *Goodness!* Did I just say that? It sounds like I got possessed by some 1950s documentary dude or something!"

The sophomore said, "No, it just sounds to me like you're growing. Yes, you have a point or two. But I still believe this esoteric stuff *is* practical for many folks if they don't overuse it or underuse it."

{87}

(Fiction Section of pages 85-88 as an
exhibit within a Nonfiction Work,
continued)

A high school junior approached them and looked and listened. The college freshman stared off into space for a few moments, considering many perspectives and angles, plus the curves of many mountain roads. The college sophomore glanced at the other two and glanced back toward the main campground.

The freshman then whispered, "Yes, I believe all of us people here at the camp *are* human beings. But now I'm wondering if some of those metaphysical theories were around *before* there were any people, animals, plants, or planets."

(Ending of Fiction Section of pages 85-88
as an exhibit within a Nonfiction Work)

Section One of Intermission One: A Biography of One of the Author's Former Neighbors, Specifically, A Man Who Reportedly Killed Himself

Sometime after Winter Storm Uri (also referred to by some as The February 13-17, 2021 North American Winter Storm) a man named Billy Mauldin met and befriended me. Although he and I never had any professional relationship with each other and did have many differences of opinion on religion and psychology, we were able to get along reasonably well most of the time for the duration of our interactions.

According to what he told various people, including myself, here is some of the background: He had worked as a professional psychiatrist prior to the 2020 pandemic. When many governments enacted major shutdowns of most business activities over extended lengths of time as a response to the COVID-19 challenges, he became forced into an involuntary severe reduction in work hours and pay. Unfortunately, he had neither learned nor practiced much accumulation of financial resources. At some stage prior to the aforementioned winter storm, he found himself living with roommates. Unfortunately for him, as a very impactful personal experience, he found out one day that his roommates stole a whole bunch of his stuff, including his driver's license and social security card. Soon he ended up in the hospital with serious health problems, then received aid from a social worker, getting to stay in a second-floor apartment unit with himself as the only official resident of that unit.

In my early meetings with him, it became clear that one of his greatest interests in life was to smoke. Also, he liked to borrow money from people, though I never lent him money. My family did help him out in another way, by giving him complimentary grocery supplies on at least a couple of occasions. That being said, there was a way that awareness spread at that apartment complex about how Billy was struggling with his life, very open to being friendly with many acquaintances, lying down in his living room with the door unlocked, and borrowing money from people.

The social worker and his relatives declined to give Billy access to having his own telephone. They also declined to give him access to his own television. He ended up having to borrow access to telephones, televisions, and Internet browsers on any occasion that he would feel the want or need to access those things.

Some of the points of disagreement between him and me were that he felt a rigid loyalty to two forms of Protestant Christianitiy and an outright hostility to the Church of Jesus Christ of Latter Day Saints (sometimes abbreviated as "LDS"), and a fixed idea that people should not think too hard about politics and religion for fear of the risk of going insane, whereas I felt differently on each of these subjects.

Regarding Christianity, I expressed a belief, largely rooted in actual experiences, partly rooted in plausible thoughts and intuitions adjacent to those experiences, and partly based on a priori realms of consciousness, that rendering unto the entirety of Christianity the amounts of reality that are properly due to Christianity would usually be somewhere between what the diverse Christian sects and subsects express about it and what Buddhism, Judaism, Islam, Hinduism, Agnosticism, Atheism, and other systems/ideologies/theologies express about it. Regarding the Church of Jesus Christ of Latter Day Saints, I expressed a belief that they have valid gateways into multitudes of realities, similar to how many other sects and subsects of Christianity have valid gateways into many realities. This is not to portray myself as having viewed a kind of absolute equivalency between all of the aforementioned groups of belief(s)-and-practice(s)-and-remainder(s); later pages in this intermission will clarify.

Next, regarding the idea about sanity versus deep thoughts on religion and politics, I expressed a belief that, in general, there are threats to sanity on both the side of thinking too hard about them and thinking too gently/softly/lightly about them, as well as any divergence from a reasonable aim toward rendering unto all religious and political presences what a given sentient being (given all of the peculiarities of that being's niche(s) within all realities) should render unto them.

About two-and-a-half-to-three months after he and I first met, he was still struggling mightily with trying to get his life back on track; he borrowed the landline phone at the apartment unit that was my then residence, near the end of the month of April 2021. On the phone with the Social Security Administration, he attempted to make progress on obtaining a replacement social security card, a follow-up attempt to other actions he had been taking. As the SSA agent and he spoke, Billy at some stage became very agitated, using what most people would consider typical curse words over the phone to complain to the SSA about how he would have to clear such a large collection of bureaucratic regulatory steps just to get the replacement card. His frustration grew and grew, eventually escalating to the point that he yelled out toward the administrative agent, "This is crazy!" Right after that, he stopped speaking and collapsed to the floor, appearing to have involuntary convulsions. Soon, I dialed 911 and spoke with the operator about the situation. Emergency Medical Service professionals arrived and tended to him, then drove away.

Time went by and I did not see Billy around. Then on

or about May 7th, 2021, a prior mutual acquaintence who had previously identified himself by the name Mario and the nickname Pancho happened to see me walk by. We spoke with each other briefly. Mario informed me that he saw Billy in a pool of blood the previous day, notified authorities, and found out that the authorities had pronounced the deeply-troubled psychiatrist dead.

A few days later, word around that apartment complex reached me that Billy Mauldin's death had been ruled a suicide by wrist slashing.

....
....
....
....
....
....
....

Regarding two individuals who are in multiple ways partially identified, though not by name, in this book:

Remember the person to whom this work sometimes refers as E92P? Besides him, there will in Section Five of Intermission One appear information about a person to whom this work sometimes confers the alias ECJ2.

....

It is unkown to me whether either of those two in-some-ways-extremely-opposite individuals ever met the Billy Mauldin described here. Also unknown to me is whether the two of them have ever met each other or know much of anything about each other as of the time of composing this.

Yet another thing unknown to me is how many, if any, people sharing the "Billy Mauldin" name combination with him the aforementioned person of that name may have met prior to evidently killing himself.

Also, I do not know with certainty whether he actually committed suicide or someone framed him as having committed suicide. Stranger still, I occasionally wonder if there might have been a third possibility, besides murder staged as a suicide and actual suicide. I heard that he had borrowed money from lots of folks, and he did speak to me of him having had an ex-wife toward whom he sometimes expressed resentment. It seemed that his main gripe against his former spouse was that he found her to be very excessively domineering in her treatment of him. All of that being said, given all facts and circumstances, from where I stand, I believe that it was most likely actually suicide, yet he probably experienced his death as definitely a suicide or as definitely a homicide, without any need to refer to it with words like, "probably" or phrases like, "most likely." It did appear to me, though, that the phone call with the SSA put together a critical mass with which to shatter his previous long-held beliefs about psychiatry.

Section Two of Intermission One: A Biography of One of the Author's Family Members, Specifically, A Man Who Reportedly Died While Asleep Soon After a Hospice Worker Gently Comforted Him While He Was In a Debilitated State, A Little While Prior to When the Hospice Worker Called the Man Who Would Go On to Become the Author of This Book to Inform Him about What Had Happened

On March 3rd, 1931, Maurice A.T. Blair was born in the state of Montana in The United States of America. His father had served in World War I in the U.S. Army. Very early in his life, his mother developed serious difficulties with diabetes, and medical technology had very limited options back then in dealing with that condition, which led to her staying in an instutition for diabetics during most of his childhood, rather than being at home with the family. For a while, his father chose to have a long-term live-in girlfriend serve as part of the household as a de facto stepmother, though not officially married into the family yet as of that time. Sometime after his mother died in December 1949, his father married the long-term live-in girlfriend, who became his second wife. His father died in December 1969, leaving multiple interdisciplinary books emphasizing the intersections of science and religion behind to be inherited by his second wife. After she died somewhat more than a decade later, M.A.T.B. inherited some of those books from her.

Growing up during the Great Depression, often in urban and suburban areas within the state of Washington, was hard. Sometimes, though his father, David Austin Blair, was a successful carpenter and a high-level Scottish Rites freemason, the family struggled mightily with finances. There were even times that David A. Blair would tell his sons (David H. Blair and Maurice A.T. Blair) that he was struggling with coming up with rent money, somehow knowing that they had sufficient money to see the family through, and demanded the money. Of the many sources that then-youngsters D.H.B. and M.A.T.B. had of obtaining it was that they would sometimes roll drunks (i.e., they would find ways to steal money from adults who were intoxicated with alcohol, often made easier in cases where the intoxicated person happened to be sleeping in public when one or both of the youngsters would be walking by). There was much street wisdom at times for them back then in the Ballard portion of Seattle, Washington, though people in plush living situations could easily question and issue snap judgments about much of the ethics or lack thereof of how they survived the economic difficulties of those times. When people jump into second-guessing and issuing rigidly-fixed judgments while having little or no comprehension of what the other person(s) actually faced, it can snowball into extra trouble for them in the long run, as reflecting and reflected by the essences of lots of ancient and modern aphorisms, many of which diametrically oppose each other to some degree or another.

At some stage, amid the difficulties, David Austin Blair snapped at Maurice Austin Theodore Blair about the idea that maybe the latter should move out and travel to Montana to live with his uncle Maurice Hoppe and his aunt Carmen Hoppe on their ranch. M.A.T. Blair, also sometimes known as Ted Blair, decided to take seriously that suggestion, which his father had not intended for him to take seriously. Ted indeed moved to Montana to work on the next phase of his life. While attending high school in Park County, Montana, residing and working on his uncle and aunt's ranch located in that county, and participating in high school team sport competitions, he became aware of some limitations and distortions from the perhaps-excessively-oriented-toward-street-wisdom mentalities he had acquired in Washington, rounding some urban and suburban perspectives with naturalistic and rural perspectives. He soon had both a set of keys to street smarts and country smarts.

Later, he would join the U.S. Army in connection with Korean escalations and various deep levels of capitalist-versus-communist escalations. During some of the early training, per his account decades later, he was among a group of men instructed to crawl carefully as an exercise, and to beware that there would be live ammunition fired above them.

Unfortunately for at least two of the other men, they stood up, apparently thinking that they would be calling the instructor's bluff. The instructor was not bluffing. They were mowed down by the bullets, and their corpses were there for the remaining trainees to see. The drill instructor said to them dispassionately something like, "See. This is what happens."

...

A while after he had in some ways followed in his father's footsteps by joining the U.S. military, he successfully completed final stages of official involvement with The Order of DeMolay upon reaching the age of 21, which some would characterize as an excellent preparation for joining freemasonry. However, he subsequently decided to avoid becoming a freemason.

...

While serving in combat in Korea, he narrowly survived, and all of the hard-won smarts and toughness from the mixtures of urban and rural environments served him well in serving America as a soldier. At some stage he decided to take steps to attempt to transfer out of the infantry and into the signal corps; he succeeded in that endeavor.

Although he briefly left and returned to the army several times over the ensuing years, and even though he had moved to the signal corps, he eventually found himself in combat again. The fighting in Vietnam was a whole other situation than what he had faced earlier in Korea. Somehow he came out of it without much physical damage, though the emotional impact of many of the experiences, both in Vietnam and back in America, was immense. He retired from the U.S. Army about seven months into the year 1972.

He decided to spend about six months hunting, fishing, hiking, and relaxing before deciding what to do next with his life. This way he could depressurize from the strange and difficult things he had been through in over twenty years of service to the army. Although he had interacted with many people and been encouraged by multiple people to consider settling down by marrying women whom they believed it could be a good idea for him to become involved with, he wished to be very careful to avoid getting into an incompatible marriage. Sometimes he used extreme, misleading, and potentially-disturbing methods to throttle societal attempts to get him to settle down. Part of this was to reduce threats to his military career, and another part was that one woman after another simply did not seem right to him as a potential wife. He was a very strong-willed and independent-minded man.

Then he started working for Raytheon as a defense

contractor. They assigned him to work in Taiwan, also known by that time as The Republic of China. There one of his coworkers - his secretary and translator, in fact - was the woman who would later become his wife and my mother.

Although his life took many twists and turns after that next phase, including much tragedy and triumph for his family at times, he had a very steady sense of duty both to reality itself (to whatever degree exhibiting through any given religion(s) and/or science(s) and/or combination(s) thereof) and to his family.

In some later years, he found himself physically threatened by two health risks: 1) after open-heart surgery about 2/3-to-3/4 the way into the year 2007 and its aftermath, side effects from prescribed blood thinners and other factors led to his skin changing into much more easily bleeding than it had previously; 2) the reason why doctors prescribed him the blood thinners, namely, an elevated risk of stroke from blood clots. After hearing from a social contact in 2010 or 2011 or thereabouts of someone in a similar predicament dying by sudden massive bleeding, he decided to go with the side of skipping the blood thinners. Although this worked well for a while, toward the end of the first half of January 2015 he suddenly had a massive stroke while resting overnight. My mother, whose name can at times take the form Ming Blair, and I found him deep asleep and unresponsive. After I called 911, Emergency Medical Services personnel arrived, were themselves unable to get him to become responsive, and determined that he would need to go to a hospital. The NICU (Neuro Intensive Care Unit) professionals informed my family that their diagnostics determined that a blood clot had traveled to his brain and triggered that stroke. A few months later, on April 26th, 2015, I received a call from a hospice worker. To the best of my memory, the male hospice worker told me that he had been gently bathing my father, who then drifted to sleep and stopped breathing, soon thereafter pronounced dead by a medical doctor.

After that debilitation yet earlier than his passing, my family had difficulties with coordinating for the possibility that he might wind up a resident of a nursing home for a period of many years, at a huge, unexpected financial burden. In relationship with that, my family hired consultant Robert Dornak to help the family with pursuing Veteran's Aid and Attendance as a monthly boost to income. Although my parents transferred a rather large amount of money to me as part of taking Dornak's advice, my family did not get close to sending in the Veteran's Aid and Attendance application, largely because the main medical doctor attending him kept leaving some required fields blank.

Earlier, over an extended period, my mother and I had decided to refrain from signing the recommended do-not-resuscitate order, hoping sincerely that giving my father every chance to somehow recover would be best. However, when things were getting near the end for him, his health deteriorated to the point that medical professionals made it clear that a hospice approach would be best and that it would be more merciful to let go of our resistance to the do-not-resuscitate order, since signing it would reduce the risk of my father having to endure the likelihood of broken ribs and other effects from extreme attempts to get him to breath again after already reaching such an extremely-weakened state. We decided to sign that order, and it was one of the many factors that led to how he became allowed to peacefully let go of the strains of this world and relax in the manner that he did, leaving this life for that great mystery that the living call death.

Section Three of Intermission One: An Autobiography of the Author of This Book, Deliberately Omitting Some Details That Were Already Revealed in Parts of His Past Books

Rather than doing this chronologically, let us delve into *pithy descriptions of much of why* I was involved with several books leading up to this one plus this very book.

Before going into a variety of other angles on these, it shall serve the reader well to visit/revisit portions of three articles that I published on LinkedIn, plus select adjacent stuff.

....

First, here is a copy of most of the text from "A Peculiar Debate from 2005 (and the Eighteenth Century) Resurfaces in 2022" (published on November 12th, 2022):

Within the conversations that Joel Edward Goza and I shared during the first four months of 2005, quite possibly the most interesting portion closely resembled the following:

Joel Edward Goza: America has been overly war-like on many occasions. Would you agree?

Maurice James Blair: Well, it could be debatable with some of the wars whether the U.S. was justified to be as involved as it was, yet I believe that our country was justified with much of the warfare.

J.E.Goza: I don't think America has been justified for much of its warfare at all.

M.J. Blair: Oh, really?

J.E. Goza: In fact, I believe that there was only one time that the United States was justified to participate in a war, and that was World War Two.

M.J. Blair: Are you freaking kidding me? If you're saying that, then that means that, among other things, you don't even believe that The United States of America was justified in its very founding with the war to separate itself from England. You actually believe that?

J.E. Goza: Yes. To me, this falls into the category of "Render unto Caesar that which is Caesar's; render unto God that which is God's." They should have found some way to accommodate the government they had while seeking a better way of life.

M.J. Blair: I think your interpretation of "what to render unto Caesar" and "what to render unto God" misses much of what the Americans were facing back then. There could be many interpretations of "Render unto Caesar that which is Caesar's, and render unto God that which is God's" that could be compatible with what the founding fathers of this country and the people who helped them did as part of conducting the revolution that resulted in this nation.

....

Next, consider text for "Glimpses of a 2002 William W. Cooper Speech" (published March 6th, 2023):

Here is a transcript of a few lines from an April 3rd, 2002 speech that Dr. William W. Cooper presented to Dr. James Deitrick's class of the Spring 2002 semester at The University of Texas at Austin, the ensuing Q&A session, and supplemental remarks:

- "Who audits the auditors?"
- "Consulting is responsible to the client; auditing is responsible to third parties."
- "Shyam Sunder ran a computer simulation that reproduced an efficient market... if people don't learn. But if people do learn, maybe markets are not as efficient."
- Upon being prompted by the author of this article to share thoughts on lists of greatest professional boxers of all time, including the question of where Joe Louis and Mike Tyson might fit in, Cooper responded, "A number of years back, researchers did a simulation, and Louis was second to Muhammad Ali, but Tyson is a great fighter."
- Soon after the question-and-answer session, Deitrick mentioned to the class that he was surprised that no one had asked Cooper to share thoughts on John Forbes Nash, Jr.

....
A detail I withheld from that article was that, to the best of my memory, my spoken question to Cooper included reference to "Lewis" (intending Lennox Lewis) without specifying which "Lewis"/"Louis," then Cooper, after hearing that sonic ambiguity, clearly specified Joe Louis in his answer without mentioning Lennox Lewis.

Here is a copy of one sentence from the article "Revealing Changes and Undercurrents while in the Beginning Stages of Bringing New Books into Our Reality" (displayed by LinkedIn as having been published on September 5th, 2022):

I am fully aware at the time of this article (to be posted late on 9/4/22 CDT / early 9/5/22 EDT), that there is much uncertainty about whether this will make much money or lose much money or both, and there is uncertainty whether my life expectancy will increase or diminish as this situation unfolds.

....
Now, consider a copy of the concluding sentence from that September article:

However, I already decided back in April 2005 that whenever I would go for bringing a new book into being, death-vs.-life, bankruptcy-vs.-solvency, etc. would be immaterial to the duty to pursue truth, facts, and realities and to give to others enhanced opportunities to pursue them as well.

....

Let us jump over to something that cuts across all ages, cultures, traditions, creeds, customs, races, ethnicities, religions, sciences, and controversies: In many respects each and every individual being has an advantage in comparison and contrast with each and every other individual belng in terms of hidden vantages and experiences, whereas in many respects each and every individual being has a disadvantage in comparison and contrast with each and every other individual being, in terms of blindspots of vantages and experiences.

....

That is a huge part of why my father would advise me, repeatedly over the span of several decades, "Think, think for yourself, don't let other people think for you." Also, he admitted to me, even when I was in elementary school, that this should apply to even any attempts he might ever make to influence my thinking.

In contrast with that, sometime circa November 1994, after gauging some of my then-tendencies as a student in a writing class that she was teaching, Dorsey Armstrong and I shared a conversation that went very much like this:

Armstrong: If someone is very open-minded and later finds out that some of the open-mindedness causes major problems, then what might be one of the options that the person should consider?

Blair: The very open-minded person might consider using the open-mindedness to determine it best to shift to being more not-quite-so-open-minded in some of the ways that could prove helpful. The person could choose to, some of the time in some of the ways, be more closed-minded.

Armstrong: Yes! That was what I was getting at.

....

My first memorable introduction to Arthur C. Clarke and his works came when my father, Maurice A.T. Blair, very early in my life, told me from memory a very brief synopsis of the Clarke short story "The Nine Billion Names of God," though my father did not mention the name of the story at the time. Years later, approximately extremely early in the twenty-first century, I finally got around to reading it. It truly impressed me then, and it has truly impressed me each time that I, on rare occasions, have revisited reading it in its entirety.

By the way, besides whatever huge and prominent controversies anyone has ever felt about December 21st, 2012 and December 12th, 2016, before, during, or after those dates, let it be known that *I read "The Nine Billion Names of God" out loud in its entirety on both of those days, once on the earlier date and once on the later date.*

....

Clarke's Fourth Law can illuminate many degrees of insight into these issues, yet I shall go ahead and relay here that Bill Wheaton as part of his September 2010 response to me on *the acc-list* mentioned the idea that multiplying that law by itself could wipe itself out, *and with it all other laws*, if someone were to take it past the outer limits.

....

Sometimes I wonder about how and why it is that many people who enjoyed the first three *Indiana Jones* movies felt uncomfortable with the fourth movie of that franchise, whereas my father and I were among those who really, truly, and greatly enjoyed all four of them [i.e., *Raiders of the Lost Ark* (1981), *Indiana Jones and the Temple of Doom* (1984), *Indiana Jones and the Last Crusade* (1989), and *Indiana Jones and the Kingdom of the Crystal Skull* (2008)].

Although I have elsewhere expressed a few theories and insights about the varying levels of popularity of those films in comparison to each other, some deeper levels may relate way beyond what any living humans might know about what occurs at the noumenal levels of the multiverse and beyond.

....

A Critical Overview of Select Challenges Leading to the Creation of This Book

I am the survivor of a child molestation incident... in which two strangers approached me circa 1982-1984 one day in Taiwan, the land of my birth in this lifetime, at a time when I was about 5 years old to 8 years old (to the best of my memory that was in the early part of that range). The female introduced the male as having a legitimate reason to hang out with me, specifically alleging he and my deaf-mute uncle to have been friends of each other; then she walked away.

Soon after she walked away, the male stranger, who appeared to be about 30-something-to-50-something years old, fondled me via direct manual pressure through my clothing. I had not yet receieved even the basic sex education of a book introduction to explaining genitals yet, and I had to deal with a then-perplexing real-life situation of a male stranger who was then multiple times my then-age grasping me in a peculiar way.

When a whole bunch of disparate lines of challenges converged on me in 1996 and I felt distracted from classwork, I chose to walk into the Counseling and Psychological Services portion of Duke University. That was about five weeks into the fall semester of that year. Multiple psychiatrists soon became involved with my life in major portions of many subsequent years. I did not find out until many years later that at a crucial juncture, before I had been given any psychiatric drug prescriptions, at least one medical professional spoke with my father behind my back about the situation, presenting two main options. One of them was to have me avoid psychiatric drugs and have in-depth conversations with psychiatric therapists to see if that could help me to get my life back on track. That would have been the much more expensive option... in terms of money. The other was to have me start taking psychiatric drugs and to have extremely limited conversations with psychiatrists: the less expensive option in terms of money.

Fully behind my back and without consulting me at all about the situation, that medical professional and my father reached a decision for me to go onto the option with the prescription psychiatric drugs. Upon hearing word of the plan to be placed on such medications (while they hid from me the option of in-depth conversations with therapists), I was mortified in that it seemed like this would be very akin to letting society play Russian roulette with my mind via the effects from adding such drugs to my system. I resisted.

Under pressure from the medical establishment, my father, and many additional sources of influence, after a while I decided to relent and let society play Russian roulette with my mind via the effects from adding such drugs to the bloodstream and neural pathways that can seriously impact my life and my ability to impact others. This led to an extraordinarily bizarre odyssey of my life going into many directions over many years, often on medications, sometimes off of medications, occasionally landing into psychiatric wards, often functioning as an outpatient. *During none of the period from when the incident of being a victim of child molestation occured until I was about thirty years old did I even find myself capable of talking to anyone about the fact of having been a victim of child molestation. I still rarely ever bring it up with anyone. Additionally, to the best of my recollection, not even once in all the years in which I was a patient of the psychiatric industry, from something like the first half of October 1996 until partway into the day May 28th,*

2021, did any of the psychiatrists and other psychiatric industry medical professionals with whom I communicated as a patient find a way to draw out of me communications with them about that fact. All of their at-times exotic diagnoses of things supposedly fundamentally wrong with me mentally totally missed that a struggle to survive in a way of overcoming trauma and confusion emanating from the childhood sudden-and-shocking-molestation-by-an-adult-stranger was a focal point of difficulties. Additionally, the way that a bunch of the events unfolded, we might not ever know if the psychiatric treatments were fully more helpful or harmful or both compared to alternate psychiatric treatments over the years that I might have otherwise received. However, I do know that after I left being a patient of that industry and for about a month-and-a-half continued with extreme struggles, I started to turn a corner, then fully turned that corner, then successfully turned one corner after another and successfully navigated many straightaways as well. I have been off of all psychiatric drugs and psychiatric treatments from the medical profession for quite a while now, and I have had reasonable confidence of coordinating my life since partway into the day July 12th, 2021.

Yes, there have been difficulties many times since then, yet in all instances there has been a core resiliency in my mind since then.

However, I often already had that core resiliency from portions of May 11th, 2011 through partway into the day September 26th, 2019, then continued it from September 26th, 2019 into early on September 27th, 2019. *That being said the psychiatric industry itself violated my core resiliency of mind early on September 27th, 2019.* A strange set of factors had very extremely endplayed me into being involuntarily placed into a psychiatric facility, where they presented me with a very long document with lots of legaleze for me to sign. I decided to take time to use a mixture of scanning, speed-reading, slow-reading, and contemplation in looking at the document, as I often do with virtually any lenghty legaleze-style contract anywhere, even though this does not conform with how most Americans glance briefly at such documents and figure that they cannot begin to fathom much or have much time to read much in said documents, simply signing quickly and hoping for the best. As I was considering things there, silently and carefully, a male psychiatric industry medical professional walked into the room.

Taking a glance at me with a hostile countenance, he soon snatched the document from me. A huge part of the reason why I had been even more careful with studying the document than I normally would with just any old random contract was that I genuinely believed then (as I also genuinely believe now) that I was not actually in need of psychiatric treatment at the time and that it behooved me then to be careful about how to manage a situation in which I had been societally endplayed into being stuck in a psychiatric ward. However, what that medical professional did at that time right after snatching the contract out from the table area in front of me was to both speak and write right there in my presence that the patient was "too psychotic to sign." Then he signed the document as a medical professional. *To be completely point-blank with him and with anyone else anywhere who might ever encounter this description from me,* whether or not there was any psychosis going on in my mind at the time - which could easily be debatable in multitudes of ways by conformists, nonconformists, psychiatrists, nonpsychiatrists, people of this or that or another faith, this or that or another nonfaith, and all persons of all remaining possible distinctions - *I WAS NOT TOO PSYCHOTIC TO SIGN. I WAS FULLY ON TRACK TO SIGN AN EXTREMELY POTENTIALLY-REALITY-INFLUENCING CONTRACT AFTER REASONABLY EXAMINING IT, within a similar length of time as considering a serious contract in some other context.*

HOWEVER, THAT MEDICAL PROFESSIONAL WHO SNATCHED THE DOCUMENT AWAY WAS NOT SUFFICIENTLY PATIENT.

Eventually, there were several key turning points in my life from July 10th-12th, 2021, and they each involved choices, decisions, and actions to sufficiently fight back against any and all beings whose actions would, in their own ways, tend to coalesce with the adult stranger's act of molesting me several decades earlier to steer reality toward the total destruction of my core mental coordination capabilities and emotional resonance centers.

Although I consider, and some of the people I have met in the many months since July 12th, 2021 have agreed with considering, that the various old diagnoses from the psychiatric industry toward me are defunct and fatally flawed as ways of pigeonholing and fixedly conceptualizing my life, health, capabilities, et cetera, I believe that everyone everywhere may possibly have at least a little capability - and at times a huge capability - while they still have any consciousness, awareness, and/or life within themselves... to overcome whatever their most primary obstacle(s) might be.

That being said, I just now used an esoteric way of managing the word "primary." You see, whatever anyone anywhere - myself included - might consider to be a primary obstacle at any given time just might - from a truer and more perfect perspective - be approaching but not quite arriving at whatever the more primary obstacle(s) might be.

....

For yet another angle on my life, please refer to the following excerpts of what I wrote by electronic mail to Evangelist Rick Hughes of Cropwell, Alabama on August 2nd, 2021:

Excerpt #1:

Thank you for the radio show "The FLOT Line." I sometimes listen to part of a show, sometimes I listen to an entire show, and sometimes I have the radio off or on a different station at the time of The FLOT Line.

Excerpt #2:

June 14th, 1976: Born in Asia, with my father being an American citizen and my mother being a citizen of Taiwan (Republic of China).

Various portions of early 1980s to mid-1994: My family sometimes attended a church that was at first named Grant Avenue Baptist Church, yet later changed its name to El Paso Chinese Baptist Church. Although my parents became members, my father became disappointed with some of the way that some of the leadership of the church did things. My father forbid me from joining it as a member.

Some time around mid-1998 until about some time December 1999: My outlook on reality often emphasized mostly what some might call Annihilationist Atheism, yet with a trace of hope for the beyond based on the possibility of highly advanced aliens.

About some time December 1999 to about some time April 12th, 2000: My outlook on reality generally emphasized Agnosticism.

Some time April 12th, 2000 to the present: I have had a multitude of outlooks on reality, changing at various times, yet focused on seeking, finding, working with, etc. truth, reality, realities, the source(s) of reality/realities, etc. Sometimes I choose one or more version of accepting Christ and Christianity, sometimes I let go of accepting Christ and Christianity, sometimes I let the acceptance of Christ and Christianity return to what I choose. I respect that various Baptists, Episcopalians, Presbyterians, Pentecostalists, Catholics, Nondenominational Christians, and others would have a vast array of reactions.

In some ways, I often choose an emphasis on Vajrayana Buddhism and Zen Buddhism, with varying degrees of these and Esoteric Christianity, Noahidism, Nondenominational Christianity, and other things, either individually or in combination, appearing, disappearing, and reappearing with the ebb and flow of statics and dynamics. It has often been this way since about May 2005.

....

On April 9th, 2023, I sent an e-mail message to *The Michael Berry Show* to inquire about why a check I sent to it via priority mail as both a birthday gift to elderly popular caller Joyce Smith and as justification for him to mention "Maurice James Blair, publisher" briefly on the airwaves (to be among the many people and businesses volunteering to be part of such arrangements of gifts to go with announcements) had not yet cleared. The USPS showed the check to have arrived well ahead of the scheduled transfer to recipient and the show indicated transfering all checks on hand to the recipient on March 24th, yet I had multiple strange experiences with delivery services in recent years, plus many bizarre experiences with people and spatiotemporal relations thus far this century; and the bank did not show the check as having cleared yet.

Though there have occasionally been legitimate criticisms of me in recent years, all too often I have had to face and push back against unjustified and illegitimate criticisms of me and outright unethical disparaging words toward me, sometimes behind my back and sometimes to my face, in recent years. The main *Michael Berry Show* e-mail address had been included on the cc line (i.e., copy concern line) of several messages in recent years in which I had used justified and legitimate criticism toward others who had attempted to unethically and illegitimately harm me out of some perception that it somehow served their interests and/or the public good. With these and many factors of how many people had interacted in recent years, I recognized that there could be great value in if the check could successfully arrive, get deposited by the intended recipient, and clear the bank. That being said, I remembered that sometimes highly successful people might choose to keep a check as a souvenir undeposited, and I felt respect toward the possibility that Ms. Smith might decide to hold the check.

For a while I was on the fence about whether or not to send a follow-up message, as much time had gone by. Suddenly, I felt inspired by Easter Sunday arriving that a carefully-arranged follow-up request could be the catalyst for good things to happen, whether that radio show might receive it with kindness or animosity. I carefully composed what I thought to be an appropriate message, yet later in the day discovered that the show, via Michael Berry's own e-mail address (though signed by his assistant Emily Bull) sent me a vicious message, even having the nerve to try to tell me to cancel the check. I responded with a combination of harsh criticism and a degree of forgiveness and kindness. After that, time went by and I noticed that the show had not thrown me out of the e-mailing list for recipients who had signed up out of an interest in the show, and I soon unsubscribed from that list. *Also, I found out that the bank listed the check as having cleared on April 10th.*

Before going into further discussion of some of this, I will show what I sent to his show during the last minute (from the perspective of Houston, TX time) of April 9th, 2023, redacting the two e-mail addresses for myself that I referenced in the text by referring in brackets to the roles they had in that set of communications.

....

Michael, Emily, et al at The Michael Berry Show,

To clarify, I was not aiming at accusing your show specifically of theft (though there was a slight suspicion of a strategic temporary delay by your show and a larger suspicion of it not ever arriving at your show's address). *I was* aiming at stating that *anyone in the chain* (including neighbors who might have intercepted it - or even - as a stretch - at least one rogue member of USPS personnel - if cognizant of portions of my 2019, 2021, and 2022 appearances on your show and influenced by any motive(s) to attempt to prevent it from arriving) if mishandling it could be causing a problem!

I do not need to cancel the check; $160 is a small

amount to keep as part of the padding with that account. So you dared to try to tell me to cancel the check; I am declining to cancel that check!!

YOU OVERREACTED!!!!!

You indicated blocking my [subscribed e-mail] address and I do not know whether you had blocked this [alternate sending e-mail] address prior to my attempting to send this message to you, yet there is a chance that you will become aware of this message even if it does not electronically arrive at your inbox, especially due to the use of the cc line and other factors.

Yes, you can choose to continue to take offense from my recent inquiry and choose to keep me blocked and think of me as a crank for daring to even bring this up with you; you have that liberty in this country, but after the several hours of caller time that you granted me (for which I am still thankful and will likely continue to be thankful for in the future, even if your show chooses to make that block permanent and refrains from future reversal of it)
spread out over time I had not expected you to read my message with such a spirit of irritability and hostility. Feel free to someday unblock me or to refrain from doing so; either way I am choosing today, to forgive you, Emily, Michael, and/or whoever else is responsible for this recent over-the-top harshness that I belicve constitutes a gross overreaction on your part.

Also, I hereby pray for your ability to discern better when someone sends you a message in a benevolent spirit without realizing that its compositional structure might otherwise trigger you, and I pray for my ability to

{115}

better anticipate how to refrain from sending a message that might trigger someone with the sensitivities that you exhibited - or to compose such a message in a way less likely to trigger someone with those sensitivities. In a future similar situation with someone else I might consider writing something more like, "That check I sent to you for you to forward to its recipient did not get deposited yet as far as I know, and it's been quite a while. Would you please confirm that you actually received it? I hope all is well with this situation."

To shine lights on any and all and every degree to which this communication breakdown reflects negatively, positively, or both on you and/or me and/or anyone else, I am choosing to include some recipients we to a degree mutually respect on the cc line (though the first three recipients are probably much more familiar with you than they are with me), plus a person I know rather well and of whom you and the first three cc line recipients were probably not previously aware.

Best Wishes to Have a Blessed Remainder of April 2023, including to whatever degrees the giving and the receiving of the rebuke of the wise may be part of this!

Maurice J. Blair

P.S., Whichever way this proceeds, I look to the past, present, and future with no regret toward our interactions and I truly believe it will become part of the good in the long run. I have let go of taking offense from this set of communications, and you can consider whether in a given minute or decade to also let go of taking offense from this set of communications.

....

Surprisingly, the Michael Berry e-mail address and the first three cc line recipients' e-mail addresses were evidently able to succeed in receiving the message, yet the fourth cc line recipient's e-mail address did not work.

That fourth intended copy-concern recipient soon received electronic communications from me via a different means on other subjects, and I became uncertain about whether or not to forward the string of back-and-forths that led to and culminated in the message eventually copied to pages 114-116. After a while I let go of forwarding it to him, though I keep the option open to someday forward it to him.

As time went by after that explosive exchange of words, I found that the show declined to give any direct response to that last message, unless the mail hosting server system were to have somehow concealed it from me or something else that bizarre were to have happened. Although I had already reduced the amount of my weekly listening to that radio show prior to when I sent the gift check payment, my respect for that Michael Berry dropped significantly in the aftermath of the conflict, and I reduced listening to him much further, to the point that nowadays I barely listen to his show, whereas in some previous years I often listened to high percentages of his radio broadcasts.

Also, for quite a while I had been on the fence about whether or not to utilize Synapsid Revelations Press Corporation to actually ever publish a book. I often considered the company, since soon after its September 2022 founding, to be similar to a network of nuclear weapons silos, only to be activated if the totality of reality would reach such a state for me to start launch sequences; yes, bear in mind that I have often considered it in very different terms as well, not always thinking of it as such a grim entity.

That being said the sudden change from *The Michael Berry Show* having been moderately favorable and kind toward me in its interactions to becoming with one assistant-signed e-mail message extremely harsh toward me and presenting a patently-false way of evaluating my life and relevancy led to my going from being on the fence about whether to create this book to making a firm decision to use Synapsid Revelations Press Corporation to publish this book with which I as the author would push back against everyone and anyone who would deserve for me to push back against them. Also, I made a firm decision to use this process to render unto a sufficient quantity of beings what should be rendered unto them (in terms of both pushing back and pulling forward, as well as relating to other metaphors of force vectors) to help uplift and enlighten those who could benefit by their encounters with it.

Section IV of Intermission One: Relating the Four Previous Portions of Intermission One to Select Questions, Answers, and Additional Questions About Portions of Reality

On Wednesday, September 11th, 2019 at 08:36:16 AM Central Daylight Time (U.S.) I sent an initial direct one-to-one electronic communication to the administrator of the website AstrumArgenteum.org the following message:

Subject: WA

Administrator or To Whom It May Concern,

When convenient, please share any comments of your choice on the following list of names:

. The Dorians (of The Dorian Invasion, circa over three thousand years ago)
. Philadelphia
. The Year 1970
. The Year 2020
. Abyssinia
. Kaspar Brunner (a Swiss clockmaker, circa over five hundred years ago)
. Kashmir

.. memory .. aware of your organization...

Regards,

Maurice James Blair

Texas

Section Five of Intermission One: Additional Focus on Some Issues Regarding The United States of America, Its Relationships, and Its Mysteries

For whatever reason(s), I never received any direct response from Astrum Argenteum officials regarding the inquiry I sent them, as reflected by the previous page.

Of course, in contrast with that, people you meet in person when traveling about will often automatically answer when they think they understand you to ask them something.

In early March, within a few days after attending the Neil deGrasse Tyson presentation from the sixth day of that month (using a ticket purchased through a secondary market for a price elevated above its initial face value), I met a svelte and enticing damsel who appeared to be in her early-to-mid 20s. I said to her something like, "May I ask if you've ever been involved with gymnastics?"

She responded with something like, "No, I'm from Colombia."

I decided not to inform her that she evidently did not understand the English word "gymnastics" in my question. Instead, I continued with other lines of conversation in a reasonable manner and ascertained that she is already married, then adjusted the conversation further to be appropriately respectful to the value of marital fidelity.

That being said, a while later, in comparing ideas, perspectives, and methods from many different religions and sciences, I could see how it is that many of them could, in their own ways help people to deal with challenges or hinder them, depending on how well people use them.

One of the medical doctors who has treated my

mother in recent months is an elderly Conservative Jewish male whom I have either only directly interacted with in 2023 (i.e., the year of completed composition of this work) or in addition to the 2023 interactions barely interacted with in previous years. Similar to how in section zero of this intermission I created an acronym for another person whom I chose to decline to name in this book as a phrase to abbreviate into the alias E92P, the medical doctor referenced in this paragraph shall here be identified by the acronym ECJ2.

After I had been reading in the waiting room one day earlier this year, probably May 9th, he and my mother emerged from the door to the hallway and medical service rooms. A friendly conversation ensued, and a ways into the conversation something about the long-ago mysterious Branch Davidian 1993 Waco tragedy influenced dynamics, on account of a magazine sitting around with a recent article revisiting it.

At some stage I said unto ECJ2 something like, "Sometimes I wonder about why it is that so many of your people are liberal. Do you have any thoughts or theories on this that you'd wish to share?"

ECJ2 responded in a manner similar to, "I think it's liberal brainwashing. They think that by siding with the liberals they are being kind and generous to people, but the effect is often the opposite."

My response to that was along the lines of, "For a while in some recent years I was very conservative. Then some conservatives started to close the gap of acting terribly toward me and showing toward others problematic behavior as well. Still, I currently tend to trend in the range from Independent Moderate to Conservative Republican. However, I think that some of it may also have to do with some liberals choosing that side out of fear. Many of them have some deep fear and find the conservatives more scary than the liberals. Some kind of association with the horrors of history, but, as I see it, there are major risks on both extremes."

Just about then his front desk receptionist lady quickly stepped aside from her duty to go to the restroom or take another type of break. I silently thought for a moment that maybe she might be the kind of liberal that her employment boss and I were speaking critically about, yet I realized then and realize now that I do not know for sure if that was under the surface of her stepping away at that juncture or if something else entirely was going on in her heart and mind and soul.

He answered with something that to the best of my memory was similar to, "There are problems with both extremes, but I think the liberal extremists are much more of a problem."

...

Something I have presented to multiple people in recent months is a metaphorical analogy between our political system and a nuclear power plant reactor system.

If things cool down too much, then it's a problem. If things heat up too much, then it's another type of problem. However, both the conservative side and the liberal side, in their own ways, act as heating things up and cooling them down. Something about the dance of all these elements and factors can result in a reasonably well-working society and economy, yet if the things get way out of whack, then things could really blow up. Clearly, back when that entire 2020 general election happened, followed by the events of the first few months of its aftermath, some stuff about the social dynamics of our political system in the U.S. very much blew up, paralleling stuff like the April 26th, 1986 U.S.S.R. Chernobyl meltdown.

I still remember how in the runup to the 2020 general election one of my former business associates, a die-hard liberal named William Crockett Walker, who also goes by the name Bill Walker, went on a scathing attack against all Trump supporters, in an effort to encourage me to vote for the Biden-Harris ticket or, at a minimum, not vote to reelect the Tump-Pence ticket.

In the aftermath of the first general election televised presidential debate in 2020, that Bill e-mailed me via Facebook Messenger a reference to the idea that "only an idiot" would vote for the Republican Party to win U.S. presidential 2020 reelection. Really? That man was outright labeling almost half of the U.S. voters idiots. Also, he layered that with playing the race card with extreme viciousness.

I tried to send him a Facebook Messenger response in which to tell him that if he actually thinks the way that he wrote, he is so brainwashed that it is almost beyond comprehension. However, the system did not allow the message to happen. Within a few days he left me an apologetic voicemail in which he said stuff like, "Politics shouldn't get in the way of friendship." Part of his word choices clearly communicated an apology to me about him "getting carried away." After a little time went by, I became able to send him a Facebook Messenger response after all, yet I sent a gentler message, emphasizing things like denouncing racism of all kinds from all sources. Also, I suggested that after he had unfriended me on that platform and sent the related messages, he and I should probably not return to friending each other on that platform.

Here is part of the backstory of stuff that led to some of that, followed by a description of some of its aftermath, which will dovetail into a more complete picture of the later blow-up of relations that happened between me and Michael Berry, as well as of many of the most volatile regions of science, religion, and politics:

In North Carolina in the mid-1990s a very liberal Duke student named Matthew Ferraguto attended many fraternity events with many of his fraternity brothers, including me, and at some point instigated a bizarre pop psychology game. He asked people to one-by-one state a favorite color, then presented to each of them some sexual personality analysis based on color selection. Basically, he presented popular cultural pigeon holes for correlating colors to personalities in what some might call a paint-by-numbers approach. Although I had for many years consistently chosen the color green to be my favorite color up until then, I stopped doing that soon after considering what happened there. Ferraguto had presented rigid stereotypes in connection with people's sexual behaviors and personalities, and I could see that no matter what anyone might choose as a favorite color, anyone using the mechanical system of pigeonholing people into personalities would wind up attempting to pin them with fixed inhibitors to a more expansive approach to effectiveness.

After that, though it was long before I became heavily involved with Jeet Kune Do and Zen approaches to life in general, I adjusted to taking a de facto Jeet Kune Do and Zen approach to the selection of favorite colors: that is, to normally not carry with me any fixed conceptualization of having any specific favorite color, then, if and when the occasion might arise to consider stating a favorite color, to, on the spot on an ad hoc basis, instantaneously choose whichever color I would believe best for that situation and occasion or to simply state not having any favorite color... or to give a long-winded explanation such as this one.

Although I dated several women at Duke, there was usually an exaggerated frigidity about me, largely because of a giant list of factors that could distract from the broader themes of this book. Part of the dating adventures, though, is extremely relevant to this discussion of the interplay between politics, religion, and science.

There were perhaps at least two different women at Duke in some overlapping years and having the name Amanda Remy. I met one of them and only many years later learned of the existence of the other one. A woman who as of 1997 was identifiable as Amanda Elizabeth Remy was in a sociology class that she and I both attended, and she indicated to me that she was a player on the women's varsity soccer team at the time that she and I went on what - by context - was clearly a date. Although she suggested that she pay for her own meal and I agreed to that without offering resistance, which by some standards (such as   according to a percentage of popular writings) could indicate it to have not been a date, and a few other factors might have suggested to some reasonably-minded judges for it to have not been a date, I showed her clearly before and during the one-to-one meal meeting significant romantic intent and she still agreed to be there with me for that. Also, at some stage in the conversation she initiated volunteering to state that she was very interested in the possibility of having a boyfriend or husband someday agree to letting her impose upon him that he could perform many of the more-typically-associated-with-housewives activities of washing clothes, washing dishes, tidying up around the house, etc.

However, the three most interesting things, in many respects, about her effect on me back then are very real and very impactful whether a given beholder chooses to judge what happened to have been a date, to have not been a date, or to have been somewhere between those two states (i.e., in a manner similar to how a quantum computer functions beyond the all-or-nothing binary way that many consider a normal computer to function):

a) When I asked her about her opinion on the then-U.S.-government-promoters-endorsed dietary guideline information, she indicated that she and other women's varsity soccer players had a very different dietary guideline that they worked with, as advised by the professionals running the team. Specifically, they would often aim toward obtaining 40% of calories from carbohydrates, 30% of calories from protein, and 30% of calories from fat; keeping bread-and-pasta types of starches down to a minimum; keeping processed sugars down to a minimum; and obtaining carbohydrates largely through fruits and vegetables rather than getting much of it from the bread, pasta, and processed sugar kinds of sources. Before that, I had learned sometime around December 1994 about the food guide pyramid approach to diet, which was often promoted by government-minded people and organizations and is geared toward a much higher percentage of calories from carbohydrates, an emphasis on including a large presence of bread and/or pasta types of starches, and a much lower percentage of calories from protein in comparison to the Duke athletic guidelines that Remy relayed to me. The food guide pyramid also espoused the idea that seriously cutting down on the consumption of fat would be helpful to reducing body fat for those who need or want to achieve that, and it included an idea that those who already had the quantity of body fat under reasonable control should feel free to consume fat as the source of a very significant perentage of calories (though still mindful to keep from having it become way too high a percentage). After meeting her and critically thinking about many factors and competing influences, I eventually often found success with using various diets that were somewhat between the then-popular food guide pyramid recommendations and what she described as expert dietary recommendations then part of the Duke women's soccer regimen. Bear in mind, though, that I have, with varying degrees of success, also used various other approaches to how to coordinate food choices in some of the years since then.

b) Although the date went well, the combined weight of everything I had experienced in life up to that time together with other factors were such that I chose to decline to initiate inviting her to a follow-up date, then time went by, then more stuff happened, then months later I decided to invite her to a follow-up date. Although I do not remember for sure the exact nature of how she turned down this possibilty, I remember with certainty how I displayed rather curbed enthusiasm toward the entirety of dating when presenting that follow-up attempt. One of the things that happened a while after I let go of the idea of seeking a second date , yet before I brought back that idea and invited her was that I walked into a library at Duke seeking helpful information on romance. I set out at the time to find the kind of information that would have motivated me to make wholesale adjustments to my life and behavior in the direction of a more exciting and active lovelife; instead, I happened to spot *Anatomy of Female Power: A Masculinist Dissection of Matriarchy* by Chinweizu. The situation, my sum-total of experiences up to that time, the totality of everything everywhere up to then, all knowns, and all unknowns combined such that I did an about-face, letting go of strong vitality aiming toward typical male adjustments, instead energizing strong vitality of aiming toward unconventional adjustments. Some of that could bring to mind the *Married... with Children* episodes in which disgruntled men would attend meetings of the National Organization of Men Against Amazonian Masterhood (i.e., NO MA'AM), though other parts could bring to mind Clarkean, Kubrickian, and Sellersian themes and memes.

I studied that book reasonably thoroughly while checking it out from that library, then checked it back in to that library. For a while I became warped in the directions of taking many of the ideas, facts, alleged facts, and perspectives of that book excessively to heart and mind and soul. Then, suddenly, very near the end of that semester, sometime around early December 1997, Amanda Remy contacted me from out of the blue to request access to a group paper that she and I and several other people worked on together for the sociology class we were taking. At that time I held the finalized version of that group paper, and Amanda wanted to borrow it in order to make a copy for her records. We met very briefly and were very friendly, though without any clear signs of rekindled romantic intent and clearly not on a brief second date. Then time went by, and I estimate that it was something like mid-January 1998 that I attempted to call her and suggest a second date, receiving either a gentle rejection spoken out loud or leaving a brief voicemail and never receiving any response (as receipt of a silent rejection).

c) As additional time went by, I came to recalibrate evaluations of comparing and contrasting things like *Anatomy of Female Power: A Masculinist Dissection of Matriarchy* by Chinweizu, various sitcom episodes dealing with dynamics between male power and female power, Internet articles, various practical direct experiences that I encountered regarding societal tendencies toward mixtures of matriarchy and patriarchy, various experiences that other people presented about societal tendencies toward mixtures of patriarchy and matriarchy, &c.

Although I had skipped voting in the 1996 U.S. general election, I chose to vote in the 2000 U.S. general election. The biggest tipping point of why I chose to vote for the Republican / George W. Bush and Dick Cheney ticket in that election was that I had the impression at the time that the balance of power between males and females in the United States of America was prior to that election almost definitely dreadfully excessively on the side of females and that I would use my right to vote as a means to attempt to shift that balance back toward either a more reasonable state or toward a better next step of pendulum motion. When the Republicans completed a controversial electoral college victory in that election I was elated.

By 2004 I was in a very different state of mind oftentimes, and I for a long time considered many advantages and disadvantages to both the Republican ticket for the presidency and the Democratic ticket for the presidency, and I had no specific societal issue as a most-important one in my evaluation at the time. I set out on a plan to wait until the last instances of time eligible before deciding for whom to cast the ballot in the general election. However, in renewing my Texas driver's license and checking the box to be automatically registered to vote, then assuming that I would be on track to be registered, I did not bother to follow up on that situation. After it became too late to register to vote, I looked around and found that I had most likely never received in the mail a voter registration card; therefore, I presumed that I was not registered to vote. Then I looked up what the processes could be and soon found out that the deadline to register to vote in the general election had passed. Therefore, I let go of planning to vote at all in the 2004 U.S. general election. That being said, when election day arrived, early in the day my mind coalesced into deciding to favor the reelection of the Bush-Cheney ticket. Without clear and distinct knowledge of whether it would make any difference or not, I then set out to telepathically project toward the U.S. populace a favorability toward the reelection of the Republican ticket for president and was delighted that Bush and Cheney won reelection as U.S. President and U.S. Vice President.

Sometime around the end of 2006 - in fact, to the best of my memory, it seems most likely to have been on December 31st, 2006 - I watched some scientific documentary about what a bunch of scientists came up with as different projected likely scenarios of how the human race will go extinct according to their models. Feeling very good about watching the episode of audio-visual explorations - very professionally presented by the way, though I do not remember what channel it was on or what the title of the show and/or episode were - I wondered from the outset what they would be ranking first on the list. Of course, as typical of such shows, they presented their list in a countdown fashion, supplying lots of discussion and details for each item before moving on to the next item. When they reached number one I was flabbergasted. Climate change?! Climate change!! That being said, I felt reasonably sold at the time that climate change was the number one threat in the long run to the survival of the human race. Early in the 2008 U.S. presidential election I decided that I would vote for the Democratic (Barrack Obama and Joe Biden in that case) ticket in that election, with the biggest tipping point for me being that I believed at the time that climate change should be the most pressing governmental policy concern.

2011 was in some ways the best year I ever had in terms of romantic interactions. One of the women with whom activities went very well for me in portions of 2011-2014 was of the name Rose Rodriguez, and she was very oriented toward a very tolerant Christian approach in which all major religions could be tolerated, many minor religions could be tolerarated, and religious liberty in general could be heavily supported. However, she also felt very strongly in favor of the Democratic Party and did not view that party as much of a threat to religious liberty. During much of that same 2011-2014 period I listened to large quantities of talk radio and watched medium quantities of mainstream television news. Comparing and contrasting a wide variety of perspectives, ranging from very conservative to very liberal, I decided for the 2012 U.S. general election to declare that the Democratic Party ticket of Obama and Biden and the Republican Party ticket of Romney and Ryan had achieved a draw with respect to my heart and mind and soul. I decided that the best poetic way to express this at that time, and to double down on the idea that governments should be concerned about both the issue of climate change and all other issues about the human impact on the environment would be to vote for the Green (Jill Stein and Cheri Honkala in that case) Party ticket. That also dovetailed with how there were many times between something like the mid-1980s and soon after the Matt Ferraguto partway-joking and partway-serious amateur psychoanalysis / pop psychology conversational game and its aftermath (sometime in the May 1995 to May 1996 time frame to the best of my memory) that I had consistently held onto having green as my favorite color.

The 2016 presidential race situation became quite a quandary for me, as I could tell that no matter which way I might vote some huge percentages of the populace could feel extreme hostility toward the choice, and because of additional factors. I decided to enact the 2004 plan of waiting until the last instant to decide for whom to vote. Eventually, I decided to forecast before going to the voting booth a high probability of voting for the Democratic Party ticket (which included Hillary Clinton as the presidential nominee) and indeed waited until being in the booth to finalize that decision. In the booth I decided to go ahead and vote for Clinton for president, and the tipping point was Trump's hideously embarrassing hot-microphone *Access Hollywood* audio recording. I thought at the time that if he really behaved in the way that he had described himself in that recording as behaving, then that would be beyond-the-pale terrible; whereas if he did not really behave in that way, then he clearly had performed an extremely terrible manifestation of impurity of speech. Either way I considered at the time the terribleness of either side of that to trump all other factors of the situation. Although the side that I voted for did not win in that election, and Donald Trump indeed ended up winning it, I felt fine upon learning of that election result.

In 2017 I found Mark Levin's talk radio presentation of information to be both very convincing and very damning toward many then-recent Democratic Party activities. Therefore, I shifted hard right. By 2020 I tended to believe that the Republican Party had the thoroughly ethical highground in general over the Democratic Party. However, I still chose to keep a slight sliver of openness to the possibility of doing something other than voting straight Republican. Eventually, I arrived in the voting booth and decided to vote straight Republican, including voting to reelect the Republican ticket of Trump-Pence.

As all hell broke loose in America in the months that followed, I mostly interpreted the facts and events of those times as a true believer Levinite, and I was outraged. I was not about to get directly involved with stuff like coordinating protests in Washington, D.C., or making a trip there to participate in such protests. That being said, at a critical juncture on January 6th, 2021, I esoterically energized consciousness itself within the universe to activate The Spirit of The Law as an entity and focused my mind on aiming that toward the U.S. Capitol to totally have its way with that capitol. My heart and mind were oriented toward a shields-off approach to what would happen next in Washington, D.C., choosing to let it be... regarding a do-what-thou-wilt nonboundary boundary toward The Spirit of The Law being able to totally have Its way with everyone and everything at the U.S. Capitol at the time.

In many ways, this is a microcosm of how I sometimes think about the relationship between the ancient Indian legends of Halahala (i.e., "The Supreme Poison" / "The Holocaust-Causing Poison") and The Twenty-Sixth Chapter of *Leviticus*.

Here is, in my own words, my recollection of the referenced ancient Indian legend: Primordial deity and/or deity-like beings decided to set out to create an elixir of enhanced vitality and longevity, yet something unexpectedly went wrong with their attempt. The resulting substance did something very much the opposite of what they had aimed for it to do: It started to annihilate anyone who came into contact with it, specifically by combustion. After much calamity, Shiva consumed the remainder of it. However, having already brought Halahala into existence, it would be virtually impossible to fully stop it from reemerging into existence and activity from time to time.

Although many could conjecture on the degree(s) of identity, if any, versus the degree(s) of nonidentity, if any, between any combination of two or more of the list YHVH, Samantabhadra, Kalachakra, Brahma, Shiva, Vishnu, and Unspecified Deity until one or more of the conjecturers might turn blue in the face, I shall decline at this instant to form any specific judgment or concept of exactly what the correct answer(s) to that mystery could and/or should be.

Next, consider a suggestion, in a similar vein to Ouspensky, Franklin, and Padmasambhava, that if people render unto any idea/behavior/being more than what is due to the idea(s) and/or behavior(s) and/or being(s), then that is tantamount to, at some higher-dimensional-consciousness level(s), an action of the mind bowing before illusions that are lacking in sight and lacking in sound and lacking in mind. Also, consider that for some people in some cases a very literal interpretation of not bowing before images carved in wood or metal or stone would serve as a great preliminary training exercise for a more advanced practice of letting go of all actions of mind bowing before illusions lacking in sight and lacking in sound and lacking in mind.

Further, at some of the more advanced levels, there can be a vast array of splitting of alternatives, two examples of which could be: 1) a realm in which some people use the observable bowing before (what appear to be physical images yet are actually) gateways into the absolute totality of reality as a means of refrainig from (at mind levels beyond physical appearances) bowing before a dogmatically-imposed excessively-simple-minded logic structure prohibiting such a method, and 2) a different realm in which some people use the refraining from that as a means of refraining from bowing before a dogmatically-imposed excessively-simple-minded logic stucture mandating such a method.

Now, let us stare directly at how a group of people involved with Jolly Old England long ago attempted to translate from source information to the English language The Twenty-Sixth Chapter of *Leviticus*, by temporarily looking away from this book in order to look at *The King James Version Bible* version of that chapter.

...

....

If you decided to decline to do that, then clearly, as evaluated by the 1913 Ouspensky expression of modeling some uses of The Tetragrammaton, your resistant force to this book's active force to influence you to do so was sufficient to render you refraining from doing so.

If, on the other hand, you decided to accept to do that, then clearly, as evaluated by that expression, this book's active force to influence you to do so was sufficient to render you choosing to do so.

(Reference: *The Symbolism of the Tarot* (1913) by P.D. Ouspensky; translated by A.L. Pogossky into English.)

....

...

In early August 2022, a few days before the thirty-sixth anniversary of the official release date of *The Transformers: The Movie*, I attended a screening of *Thor: Love and Thunder* at a Cinemark location. While there, I happened to meet for the first time a woman named Shane Yelverton, who indicated that she is the person of that name who was one of the few employees of Enron who volunteered to speak with news reporters while a changing state of play for its early 21st-century debacle was a hot news story.

Early in the interactions we were able to get along reasonably well it seemed. However, she repeatedly pressed for the idea that the only correct relationship any human being can have with reality would require as part of it a complete adherence to accepting Christianity as the only true religion and rejecting all other religions as being false religions. I mentioned to her a variety of things in the same and in similar veins as the August 2021 e-mail message that I sent to Rick Hughes (refer to pages 111-112 if necessary).

Nevertheless, she would occasionally press me about this issue again and again by phone calls and text messages. Finally, she seemed to completely cease making such attempts (and ceasing to initiate any attempt at new rounds of one-to-one communications with me, while I have also ceased to initiate any attempts at new rounds of one-to-one communications with her) after:

1) I felt insulted by how she concluded her October 1st, 2022 at 6:58 PM Houston time text message to me with, "I feel sorry for you because you are spiritual [sic] lost. Jesus said. [sic] Nobody cometh to the father but by me. All other religions do not have a risen savior. I've told u [sic] the truth. That's all that I can do."

{The "[sic]" label above is indicative of variances from generally accepted grammar, punctuation, et cetera, in coordinating that composition; specifically, that she used "spiritual" where conventional grammar would have indicated the word "spiritually," that she placed a period instead of a comma after "Jesus said" in the process of presenting one of the variations of quoting

"John 14:6," and the oft-used texting "u" for "you."}

2) Although I usually keep many of my deeper capabilities hidden from almost everyone almost all the time, I decided to really let her have it with bluntness, spirituality, and telling her what I suddenly believed she demonstrated a need to know, revealing to her several alternate perspectives point blank. On October 2nd, 2022 at 7:05 PM Houston time I sent her a text message stating, "To be quite honest with you, I believe that I often have one of the truest versions of Christianity, perhaps even the truest version of them all: Esoteric Christianity with an emphasis on Tertium Organum and things adjacent to that Ouspensky book. That also allows letting go of Christianity sometimes and to some degrees and using endless varieties of syncretic religion(s). You can cling to your petty tribalist approaches to your overly-simplified concept(s) of Christianity and the rest of everything, but I truly believe that you and I and everyone else sooner or later will have future tests in which the only true way to be attuned to all realities involving Jesus and all other beings will be to go further than we had before in terms of perspectives like Matthew 6:33. Next, regarding the couch at my house: unless and until many radical changes were to happen with multiple things, the house I co-own will probably not be available for Tony Holubik to stay at overnight, though he can return to visit it briefly during the day; he has not stayed there overnight before, and I am not so sure that it will in the future be a good idea to let him stay overnight, though there might be an outside chance of it someday changing. Third, as a reminder in connection with Deuteronomy 13, just because some folks think they know that someone or something unfamiliar to them is clearly in no way, shape, or form a manifestation of the God of their fathers, it does not always mean that they

are right in their assessment. Think about that, and if any of your ministers tell you in reaction that they believe they are absolutely right to condemn what I presented to you in this text, then maybe they should sometimes think and pray about how much they are certain or uncertain of about the full mysteries of the holy spirit, and whether they and you and everyone else might in the future be more aware of all the truths and realities in and adjacent to this text message response. Anyone who judges another to be lost based on superficial conceptualizations spoon-fed by simpleton ideas of reality might temporarily be able to use that like a tricylcle or a bicycle with training wheels, but as the person becomes more spiritually mature and capable and aware, he or she may let go of that as part of seeking and finding greater levels of awareness of God, righteousness, truth, etc."

{There are various ideas about whether or not to italicize titles of books viewed as within larger books.}

Please consider how, similar to the notion that Clarke's Fourth Law squared could dissolve itself and all other laws, some mind patterns could express the notion that a combination of *Genesis* 1:26, *Exodus* 20:3, and *John* 9:39 can be tantamount to *The Holy Bible* dissolving itself into utter Zen transcendence.

Back from this segue to the previous emphasis on politics: Eventually, I let go of feeling quite as certain about Mark R. Levin's valuations of the Republican-Party-versus-Democrat-Party dynamic in this country in recent years being about 85% to 99.7% true.

Rather, as time went by, especially since about the middle of 2022 I have often found myself returning to somewhere between the Mark R. Levin part of the political spectrum and the Maureen Dowd part of the political spectrum.

A huge part of this is how Liza Darnton's acceptance

of me for LinkedIn networking purposes after I sent her an invitation to join each other's networks was the decisive event in a chain reaction of long-term healing, with April 7th, 1995 to June 19th, 2022 amounting (in many ways) to her having perfectly interacted with me for the span from when she kissed me repeatedly in a hyper-rapid fashion at the conclusion of the first and only date that she and I have had with each other (with those being the only clear and distinct physical displays of affection we have shared) to the time that I found out that she had accepted my networking invitation.

Although there are some ways that True Buddha School Founder Lu Sheng-yen, my father Maurice A.T. Blair, the Jessica Trend who dated me on May 11th, 2011, and a few other people I met and learned from were also perfect in their interaction with me, in the regular sense the referenced Liza Darnton and I together achieved something probably seldom witnessed on Planet Earth: After a confluence of events were such that she became the first woman I truly dated and the first woman I truly reached base with, she and I decades later accepted each other, the past interactions, and the new networking opportunity, and that freed me from the often-typical adult human imprint of a lingering first highly-charged romantic loss, converted to a win. In many ways, though I have experienced many romantic losses and few romantic successes, *there was a super win* (even with how that Liza went silent in response to follow-up attempts from me to start one-to-one e-mail dialogues after accepting the joining of networks, and even with how Spring Semester 1995 is most likely going to wind up the only period in this lifetime of our meeting each other in person). I can listen to references to lingering anguish from first romantic loss in popular songs and find myself with the grass greener on my side of the fence.

****

Two additional persons who were very helpful were Whit Cobb, an elderly coworker with me during some of my gym-monitoring work study at Duke Health, Physical Education, and Recreation, and Amin Nosrat, who delivered an April 19th, 2001 exit interview for the internship that I completed on that date.

Although the set-to-begin-in-September-2002 full-time/regular position with Arthur Andersen did not materialize, mainly due to the 2001-2002 auditing debalces that it faced, Nosrat, in seeking to help me to improve my professionalism, inadvertently caused me to partially merge with replicating elements of multiple characters from the movie *Dr. Strangelove* (1964), especially in ways that would dovetail with multiple tantric practices, esotericism, and the occult. A huge part of this was how I shifted to a total letting go of the normal human cycle of dissatisfaction to satisfaction to next dissatisfaction to next satisfaction, rarely subsequently allowing that cycle to return, going over to an escalating and de-escalating perpetuity as a much more primary basis of vitality.

That businessman Nosrat was probably oblivious to how his constructive criticism (of repeated, occasional, general goofiness in speech and conduct out-of-step with his ideal of effective white collar interpersonal behavior at Andersen) aimed toward my improvement as a business professional led to my immediately for a while losing all interest in great swathes of physical pleasure, discovering heightened spiritual awareness with which to parlay previous spiritual advances into further advancement, and coordinating pleasure with spirituality in a highly integrative fashion.

One example is that since the time of the constructive

criticism of that April 19th, 2001 interview to the time of publication of this book, I have completely avoided intentional orgasm, even when engaged with intense romantic physical intimacy with partners (in accordance with some ranges of Vajrayana Buddhism, Esoteric Christianity, and other spiritualities converging with a somewhat Sramanist curbing of pleasure / a rather Kubrick-Cotton-Ripper-esque sexual ethic), and will most likely continue this restriction permanently unless and until radical changes were to take place.

One such change could be if I were to marry a woman and find the marriage to be developing in a way that would be spiritually correct to go with such a reversal, for example to increase the likelihood of procreation. Another such change could be if I were to get very involved with some of the alternate tantric practices in which such a reversal would be part of the very religious rite formulations (as a set of examples, consider some of the extreme heterodoxy described in tantric works involving Hevajra, if you dare).

Recommended Further Reading:

- *The Book of Buddhas* by Eva Rudy Jansen
- *The Seven Habits of Highly Effective People* by Stephen R. Covey
- *The Dhammapada* as translated by Ananda Maitreya
- "All the Time in the World" by Arthur C. Clarke
- *Of Mice and Men* by John Steinbeck
- *The Subtle Knife* by Philip Pullman
- *Fahrenheit 451* by Ray Bradbury

Chapter 9: A BUSINESS TAROT DECK POSSIBILITY

0. Conflict
XXI. The Economy
XX. Profitability
XIX. Confidence
XVIII. Uncertainty
XVII. Goals
XVI. Crisis
XV. Deception
XIV. Action
XIII. Bankruptcy
XII. Sacrifice

I. Chemistry
II. Entrepreneurship
III. Chief Financial Officer
IV. Chief Executive Officer
V. Business Philosophy
VI. Salesmanship
VII. Competition
VIII. Truth
IX. Reflection
X. Solvency
XI. Motivation

Ace, Deuce, 3, 4, 5, 6, 7, 8, 9, 10, Worker, Supervisor, Vice President, and President of:

- Operations & Management
- Marketing
- Finance & Accounting
- Public Relations

XXII. Pendulum of War and Peace
XXIII. Harmony
XXIV. Authenticity
XXV. Movers
XXVI. Covenants and Removers
XXVII. Effectiveness

Chapter 10: SYMBOLIZING PORTIONS OF THE 471 MILLION B.C.E. TO 1999 C.E. PERIOD USING ALL FIVE OF THE AFOREMENTIONED DECKS AND COMBINATIONS THEREOF

"471000000-4005 B.C.E."

II. Duality

Ace of Preservation

XXI. The Great Ultimate
XXI. The Economy
VI. Passion

Ace of Destruction

XXVII. Effectiveness

Ace of Commerce

XX. Profitability
V. Religion
V. Business Philosophy

XIII. Endings

"4004 B.C.E."

Ace of Creation

Tetragram Models

I. Unity   II. Duality

Voidness Models   0. Nonduality &/or Plurality

2 of Creation	3 of Creation	
4 of Creation	5 of Creation	IX. #
7 of Creation	6 of Creation	IX. ~
9 of Creation	10 of Creation	XIV. #
8 of Creation	XV. Deception	
VIII. Truth	Ace of Destruction	

VIII. Super Symmetry   XV. Super Chaos
VII. Pursuit   XVI. Law

XI. Motivation   XII. Sacrifice

"Highlights from the Merging and Separating of Ideas Across Sciences and Philosophies in the 5 B.C.E. to 1550 C.E. Period, including Modern Metaphors for Those Pre-Modern Times"

I. Unity

II. Entrepreneurship

II. Duality

XX. Moment of Truth

III. Entrepreneurship

IV. CEO

III. CFO

XIX. Confidence

XX. Profitability

III. Methodology

IV. That of All Methods and No Methods   XIX. Execution

VIII. Truth

XV. Deception

2 of Creation

2 of Destruction

Tetragram Models

Voidness Models

"Emergence of Protestants and the Expansion of Diversified Christianity During the Early Protestant Reformation"

VIII. Truth	XV. Deception
2 of Creation	2 of Destruction
3 of Creation	4 of Destruction
5 of Creation	6 of Destruction
7 of Creation	7 of Destruction
7 of Revelation	9 of Revelation

!?   ?!   !!

Addtional VIII. Truth

"Protestantism vs. Roman Catholicism vs. All Else in Portions of Europe and the Middle East in the 1600-1898 Period"

VII. Competition

XVI. Crisis

Ace of Creation

Ace of Destruction

Ace of Commerce

Ace of Preservation

Ace of Revelation

!?   ?!   !!

"A Snapshot of the 191 Million B.C.E. to 1949 C.E. Period From Multiple Elevations and Locations Ranging From Venus to Earth to The Earth's Moon to Mars to Jupiter"

One Complete Deck Each of:
^{235}U Tarot
^{239}Pu Tarot
^{3}H Tarot
^{2}H Tarot

One Supplemental Complete Suit of Revelation

XVI. Crisis   VIII. Truth

VII. Competition   XIV. Action

IX. Reflection   0. Conflict

I. Chemistry   XVIII. Uncertainty

XVII. Goals   XI. Motivation

XII. Sacrifice

9 of Commerce

5 of Commerce

Ace of Destruction

XIII. Endings

Ace of Commerce

7 of Commerce

5 of Marketing

ZX   MMM   XYZYX

3 of Commerce

VI. Passion

10 of Creation

X. Beginnings

? !? ?! !!

A Complete Business Tarot Deck

A Complete Deuterium Tarot Deck

A Full Suit of Commerce

A Full Suit of Creation

A Full Suit of Destruction

!

"Pentecostals, Baptists, Shakers, Orthodox Jews, Reform Jews, Shiite Muslims, Sunni Muslims, Nondenominational Atheists, Nondenominational Christians, Nondenominational Jews, and Agnostics Compete for Influence in Portions of the Middle East, Northern Africa, South America, and North America in Portions of 1946-1996"

5 of Commerce
10 of Revelation
4 of Destruction
5 of Destruction
XVI. Crisis
X. Solvency
VIII. Truth
9 of Destruction
XV. Deception
Ace of Destruction
10 of Operations & Management
9 of Operations & Management
8 of Operations & Management
7 of Operations & Management
6 of Operations & Management
I. Unity
XV. Super Chaos
II. Duality
XY XX GMG RRSS SSRR
2 of Destruction
3 of Destruction
10 of Commerce
XIII. Bankruptcy

"Ian(s), Johann(s), Lee(s), Diana(s), Extravert(s), Introvert(s), Ambivert(s), Santana(s), Madonna(s), Prince(s), Jimi(s). James(es), William(s), Williams(es), John(s) Robert(s), Roberts(es), Charles(es), Oliver(s), Jerry(s), Garry(s), Merry(s), Mary(s), Matthew(s), Matthews(es), Gary(s), Gerry(s), Salamander(s), Gerrymander(s), Vase(s), Gervaise(s), Bid(s), Bad(s), Good(s), Baghdad(s), Burma(s), Turkey(s), Constantine(s), Morbia(s), Westmoreland(s), Chan(s), Zen(s), Them(s), and You(s), 431 B.C.E.-1999 C.E. / 431 B.C.-1999 A.D. / 431 C.E.-1999 C.E. / 1999 B.C.E.-431 C.E. / 1999 B.C.-431 A.D. / 431-1999 Ambiguated"

X. Beginnings
Ace of Destruction
Ace of Creation
XIII. Endings
2,999 Complete ^{235}U Decks
3,558 Complete ^{239}Pu Decks
I. Unity
0. Nonduality &/or Plurality
XV. Super Chaos
XVI. Law
II. Duality
4,000 Complete ^{3}H Decks
4,500,000 Complete ^{2}H Decks

90 Million More Instances of A Complete Suit of Creation Paired with A Complete Suit of Destruction

!   !!   !?

V. Religion   ?

XVIII. Unknown

XVII. Excavation

XYZYX

EVE

ADAM

VEV

MADA

President of Public Relations

President of Marketing

WLW

GMG

?!

?   !   !?   ?!   !!

Conquerer of Commerce

Conquerer of Revelation

Conquerer of Destruction

Conquerer of Preservation

Conquerer of Creation

10 of Revelation

10 of Creation

10 of Destruction

10 of Preservation

10 of Commerce

9 of Revelation

10 of Creation

10 of Commerce

10 of Preservation

10 of Destruction

10 of Revelation

Ace of Creation

Ace of Destruction

Ace of Commerce

Ace of Preservation

Ace of Revelation

!?   ?!   !!

XXVII. Effectiveness

20 Complete ^{235}U Decks & 21 Complete ^{239}Pu Decks

1,880 Extra Complete Suits of Commerce

VIII. Truth

1,900 Extra Instances of VI. Passion

?   !   !?   ?!   !!

Conquerer of Commerce

Conquerer of Revelation

Conquerer of Destruction

Conquerer of Preservation

Conquerer of Creation

!?   ?!   !!

200 Complete ^{235}U Decks & 201 Complete ^{239}Pu Decks

180 Extra Complete Suits of Commerce

199 Extra Instances of VI. Passion

VIII. Truth   XVI. Crisis

XX. Profitability   XI. The Economy

Additional Instance of a Pairing of:
XX. Moment of Truth   XI. The Great Ultimate

A Reiteration of: XVI. Crisis   VIII. Truth

XV. Deception   VIII. Truth

X. Beginnings
Ace of Destruction
Ace of Creation
XIII. Endings

I. Chemistry   I. Unity

VIII. Truth   XV. Deception

2 of Creation   2 of Destruction

Supervisor of Operations & Management

Vice President of Finance & Accounting

Conquerer of Creation

Conquerer of Destruction

Conquerer of Preservation

XV. Deception   VIII. Truth

Ace of Operations & Management

5 of Commerce

4 of Preservation

3 of Commerce

2 of Preservation

Ace of Commerce

2 of Commerce

3 of Revelation

4 of Destruction   5 of Creation

6 of Revelation

7 of Commerce

8 of Creation

9 of Commerce

10 of Creation   10 of Destruction

!

?   !?   ?!

!!

"The Rise and Fall of the Manichean Religion, and Its Partial Rebirth in Portions of Later Syncretic Religions Such as Yiguangdao and in the emergence of George Lucas finding inspiration for *Star Wars: A New Hope*, Reflected by Portions of the 216-1977 Period of Time"

7 of Commerce

9 of Commerce

5 of Commerce

2 of Preservation

II. Duality

5. Religion

XXI. The Great Ultimate

XIII. Endings

XII. Ancient Wisdom

2 of Destruction

3 of Creation

XVII. Excavation

XVIII. Unknown

Ace of Destruction

Ace of Creation

II. Entrepreneurship

V. Business Philosophy

XIX. Confidence

XYZYX

"The 1440s versus The 1940s"

I. Unity

II. Entrepreneurship

III. Methodology

V. Religion

XX. Moment of Truth

XXI. The Great Ultimate

XYZYX

DTSP

PTS

Ace of Creation

Ace of Destruction

I. Chemistry

0. Conflict

V. Religion

GMG

PTSD

STP

Ace of Creation

Ace of Destruction

VI. Passion

VII. Pursuit

II. Duality

"*Revolver*; *The Good, the Bad and the Ugly*; 1966; *Life is Beautiful*; 1997; Issues of The Eye of the Beholder, The Eyes of the Beholders, The Minds of the Beholders, The Hearts of the Beholders, Robert B. Cialdini's *Influence: The Psychology of Persuasion*, Dale Carnegie's *How to Win Friends and Influence People*, 1965-1999 in general, George Strait's *Lead On*, Dr. Dre's *The Chronic*, Bob Dylan's *Bringing It All Back Home*, perceptions and/or realities of the Handsome versus the Homely, and Chris Carter's 1996-1999 television series *Millennium*"

201 Complete ^{235}U Decks & 201 Complete ^{239}Pu Decks

188 Extra Complete Suits of Commerce

919 Extra Instances of VI. Passion

XXII. Pendulum of War and Peace

XXIII. Harmony

XXIV. Authenticity

XXV. Movers

XXVI. Covenants and Removers

"Rock, Pop, Soul, R&B, Rap, Hip Hop, Country, Jazz, Folk, Classical, World, New Age, Old Age, and/or Hybrid Music of the 1967-1999 Period"

One Complete ^{235}U Tarot Deck

Five Complete ^{239}Pu Tarot Decks

Two Complete ^{3}H Tarot Decks

Three Complete ^{2}H Tarot Decks

XXII. Pendulum of War and Peace

XXIV. Authenticity

XVIII. Uncertainty   V. Business Philosophy

XVII. Goals   VI. Salesmanship

XXIII. Harmony

I. Chemistry

0. Conflict

XIX. Confidence

"Of Maritime Oceanliners, Military Submarines, Civilian Submarines, Icebergs, Spacecraft, and People Under Pressure, 1830-1999"

Complete Suit of Preservation   Complete Suit of Destruction
Complete Suit of Creation   Complete Suit of Commerce
II. Entrepreneurship   X. Solvency   XIII. Bankruptcy

V. Business Philosophy   XVI. Crisis

VII. Competition   I. Unity   II. Duality
XV. Super Chaos   VIII. Super Symmetry
XYZYX   XX   XY   YX   GGG
STP   PTSD   DSTP   PTS

RRSS   SSRR

XVIII. Excavation   VII. Passion

X. Beginnings   XXI. The Great Ultimate

XXI. The Economy   XIII. Endings

XI. Modern Wisdom   XII. Ancient Wisdom

GMG   TTT

XXII. Pendulum of War and Peace

XXIV. Authenticity
XXV. Movers

XXVI. Covenants and Removers

XXVII. Effectiveness

BONUSES BETWEEN CHAPTERS TEN AND ELEVEN

Part I: By the way, the aforementioned Robert Dornak (ref. p. 99) was the consultant whose name the first sentence of the preface withheld (ref. p. v).

Part II: Let it be known to readers that I, Maurice James Blair, chose in January 2023 to send a gift of one paperback copy each of Katsuki Sekida's *Zen Training: Methods and Philosophy* and Maurice James Blair's *All Things under and over the Sun and Stars: Enigmas in Various Stages* to then-Missouri-Correctional-System-death-row inmate Leonard Taylor. I called ahead of time in order to determine the proper procedure and discovered that their system would have to evaluate any book gift for suitability before deciding whether to ban a book or multiple books or to transfer the gift to its intended recipient.

Although it was unknown to me then and remains unknown to me now to what degree he might have in actuality been innocent or guilty, I believed it best, based on all factors, intuitions, and sensations known and unknown, to go ahead and arrange to send that gift via an online retailer. On May 26th, 2023, I chose to make a follow-up call to the facility at Mineral Point, Missouri, to find out whether the Missouri Correctional System had deemed the gift acceptable to go to the prisoner after its delivery. The mail room supervisor confirmed that both books were accepted by the system and not banned; however, she indicated that they make such determinations for each book each time that someone attempts to send a copy of it to a prisoner in their system, and any past instance(s) of a given book passing their evaluation to be suitable would not guarantee its passing again in the future.

Part III: Here is a copy of the first outtake for the descriptive text to go with the back cover of this book:

Select reactions to previous works by Maurice James Blair and some of how this work relates:

When he contacted the U.S. Secret Service by letter to inform it that he would be arranging to send six copies of *The Dimetrodons, the Dorians, and the Modern World*, to attempt to give one copy each to Presidents Biden, Trump, Obama, Bush, and Clinton pending that service's evaluation of its suitability and one copy to the Secret Service itself to study, as a gift of insights following up on challenging realities, someone in the chain of custody marked the package "RTS." Informed that the printing office he paid to print and send those copies had received the unopened package per the preemptive "return to sender" rejection, he e-mailed the printer that he would gladly become the replacement recipient, to then bide time to decide if and when to select other people to whom to give those.

After he arranged to send one copy each of *All Things under and over the Sun and Stars: Enigmas in Various Stages* to two congressional committees as gifts directly from a different printer, together with a note about how it was at the time the best he could do to use about 68,500 words to convey the truth, he received no direct response, yet he was encouraged by how they accepted that gift without preemptively returning the packages to sender.

Several recipients of *The Dimetrodons, the Dorians, and the Modern World: Revised Edition* and several recipients of the two aforementioned books have reacted with silence; others have voiced mixed reactions. Though readers could feel wonderful or terrible or both about the exotic arrangements and open-endedness of those novels, beings who study at least two out of those three fiction books combined with the present nonfiction book *Science, Religion, Politics, and Cards* may eventually wield awareness and knowledge that they never previously dreamed possible.

Part IV: Please bear in mind:

- In 1965, The Byrds covered the Pete Seeger song "Turn! Turn! Turn!" (which in turn had used English translation of select portions of the third chapter of the book that from *Tanakh* perspectives is sometimes referred to as *Keholet* and at other times referred to as *Ecclesiastes* and from *Biblical* perspectives is often referred to as *Ecclesiastes*).

- In 1965, according to records, Philosopher Joel Feinberg presented to the public the first in a series of editions of the anthological philosophical book *Reason and Responsibility*. Decades later, he would choose to include a modified and limited presence of Philosopher Paul Churchland in at least one edition of one of those books, and it was via Feinberg forwarding Churchland expressions in part of a textbook that Philosopher Tad Schmaltz then presented to various of his Fall Semester 1994 students, including myself, an introduction to Paul Churchland and a mostly motley assortment of philosophers competing with each other and everyone else for what in the world (or out of this world) that people could and/or should and/or would cognize.

- Some might consider "motly" to be a valid way to contract the compound adjective "mostly motley," whereas others might consider someone daring to ever present such a thing (intentionally or not) as instantaneously deserving the death penalty/etc.

CHAPTER ELEVEN: INTERMISSION TWO

Although Covey's *The Seven Habits of Highly Effective People* mentioned the value of having principles at the center of one's life, there is the factor of how principles often coexist in opposites.

One person can have a niche in which many principles coalesce into a coherent and effective way of life unless and until the situation changes dramatically, and another person can have a niche seemingly alien to the niche of the first person. Also, an advanced practitioner of esoteric use(s) of religion(s) in general, whether nominally of this or that religion or combination of religions, can have very adaptable and everchanging uses of idea structures.

An example is "less is more" versus "more is more." Here are more examples, more or less:

- "Best to play it safe" versus "Best to seize the day"
- "You can catch more bees with honey than with vinegar..." versus "Nice guys finish last..."
- "Better to be nice than to be mean..." versus "Better to be mean than to be nice..."
- "Always do what is right..." versus "Sticking too tightly to your idea of only doing what you believe to be right could result in you acting as too much of a goody-two-shoes type of person..."
- The sixteenth and seventeenth verses of the seventh chapter of *Ecclesiastes* versus a given person's perception of those two verses versus another person's perception of those two verses versus yet another person's perception of those verses...

Two of the outtake ideas for the title of this book were "Verses Versus Verses" and "If You Think That Swing Voters Should Swing From a Tree, then Maybe Our Country 'tis Not of Thee."

Another example of principles versus competing principles is to consider one of Thomas Jefferson's most famous warnings and contrast it with its opposite.

- "I, however, place economy among the first and most important republican virtues, and public debt as the greatest of the dangers to be feared." - Thomas Jefferson to William Plumer, July 21, 1816 (source: https://www.loc.gov/collections/thomas-jefferson-papers/articles-and-essays/selected-quotations-from-the-thomas-jefferson-papers/ as accessed on June 29th, 2023)

- "I, however, place economy among the last and least important democratic virtues, and public debt as the least of the dangers to be feared." - Opposite of Thomas Jefferson implicitly in the minds of reality as an accompaniment to William Plumer, July 21, 1816 (source: the processing to create this bullet point happened by multiplying Clarke's Fourth Law several portions of the quotation that preceded the bullet; since it is so closely adjacent to the public domain quotation that it shares isometric opposition with it, it may be regarded as a public domain meme that had 1816 simultaneous co-creation in the public sphere, though seldom - if ever - daring to explicitly appear in print until appearing here)

Although the first statement explicitly happened way back when, and it is unknown to the author when the second statement first explicitly happened (including whether or not anyone previously dared to put it into print prior to its inclusion in this book), there is a primal scream about the pair.

In much of the modern times, liberalism in America has had a tendency to conceive of reality as one in which the more natural and free and open use of choices by individuals tends to trample under foot the disadvantaged, leaving them little or no room to thrive. In contrast, in much of the modern times, conservatism in America has had a tendency to conceive of reality as one in which the more natural and free and open use of choices by individuals tends to open doors of opportunity to the disadvantaged, letting them thrive or fail to thrive depending on their own merits and the mysteries of fortune and divinity.

At some core levels, I believe that reality can exhibit in multitudinous cases multitudinous examples of supporting every portion of the full spectrum between both extremes. If Person A's, Person B's, and Person C's experiences, concepts, intuitions, and feelings are very different, then it could make sense for them to have three very different sets of voting behaviors. However, if any one of them becomes fixated on the idea that only one portion of the political spectrum has any truth to it at all, then that person could easily become vulnerabe to propogandists who heighten that fixation and energize a willingness to lash out at people who refuse to conform to voting exactly the way that he or she would prefer people to vote.

If a similar mentality is taken in the direction of classifying all manner of spriritual and lifestyle choices as all-or-nothing totally correct or totally incorrect, then the given person and others drawn into a similar mindset could start to become capable of virtually anything in terms of a willingness to destroy the lives of dissidents. What tends to lead to this could vary each time it crops up in history, but the most underlying of causes is, arguably, objectively unknown to everyone. Sure, maybe a given set of theories, beliefs, and experiences can produce all manner of people thinking that they have a reasonably complete and truthful understanding of what caused the most shocking atrocities of history, yet every time a person reaches a new horizon of such understanding and insight, there are further horizons available. (Credit Ancient Wisdom.)

Although in some of the remainder I shall be showing messages composed by other people and sent to me, doing this without having sought permission from them to include them in this book, I believe that within the context of analyzing how the illustrative examples of correspondence happened and information adjacent is well within the fair use of copyrighted materials. These will help clarify the preceding portions of this book.

Here is a copy of September 11, 2020 to October 10, 2020 Facebook Messenger correspondence between former coworker Willam Crockett Walker and me (using Houston time demarcations) (including all grammatical, punctuation, etc. divergences as-is to the best of ability, declining to add "[sic]" for missing punctuation, etc., and mostly declining to add explanations/notes):

9/11/2020 at 10:04 AM, Crockett Walker sent:
"ok sounds good let me know what you think im good for either"

9/15/2020 at 3:59 PM, Maurice James Blair sent:
"Sunday the 27th. It might turn out to be 10 AM, though, as I am uncertain ahead of time about what the timing of breakfast will be."

9/22/2020 at 8:18 AM, Walker sent:
"Ok sounds good"

9/28/2020 at 8:10 PM, Walker sent:
"Hey it was good to talk to you. I always enjoy catching up"

9/30/2020 at 10:02 AM, Blair sent:
"Yes, it was good to talk with you. I enjoyed meeting each person I met at the gathering."

10/2/2020 at 8:53 AM, Crockett Waker sent:
"Hi, Maurice, I really dont want to even bring this up, but I feel I have to. I just want to ask you if you know that the person you are considering supporting is supported fervently by David Duke, the KKK and the Proud Boys? I do not think you are a racist, but they are groups that used to hang black people from trees only 75 years ago. And they LOVE donald trump. Are you in fact a white supremacist or racist? I have to ask you because trump certainly is and if you support him I'm not sure."   [Note: I, M.J.B., skipped Facebook for several days during around that time.]

10/5/2020 at 3:19 PM, William Crockett Walker sent:
"Maurice, I'm sorry but if you support Trump, you are supporting the KKK, not to mention a complete idiot. We cannot be friends anymore. If you come to your senses please text me otherwise good luck joining the KKK"   [Note: W.C.W. skipped placing a period.]

[Note: I, Maurice James Blair, the author of this book, have never joined the Ku Klux Klan.]

10/9/2020 at 6:05 PM, William Crockett Walker sent:
"Hey man I sent you a voice mail but I want to say Im sorry for the rants I wrote above this message. Please forgive me. Politics should have nothing to do with our friendship and you and your mother are welcome at our house anytime"

10/10/2020 at 9:08 AM, Maurice James Blair sent:
"Thank you, Bill. I disavow racism in all its forms, yet believe there is a degree of it in many places. Yesterday I sent a brief philosophical e-mail to you."

10/10/2020 at 10:31 PM, William Crockett Walker sent:
"Thanks man! I really do appreciate your friendship and admire you for who you are"
[Note: W.C. Walker again skipped placing a period.]

On October 9th, 2020, via an e-mail transmission away from Facebook, I sent to Bill Walker (i.e., the aforementioned William Crockett Walker) the following message, with "perspectives" as the subject:

Bill,

After receiving your recent voicemail and comparing and contrasting it with recent Facebook communications, here are a few perspectives.

Here are four contrasting statements:

I. "All idea structures have a degree of reality, and all idea structures have a degree of illusion."
II. "Not all idea structures have a degree of reality, and not all idea structures have a degree of illusion."
III. "Idea structures transcend the concepts of 'reality' and 'illusion.'"
IV. " "

I agree with the idea in your voicemail message that we could restore being friends in a significant sense, though with caution with respect to the issue of politics. That being said, I believe it is probably best that we refrain from bringing back the Facebook friending of each other.

Best wishes with the well-being of sentient beings of all kinds everywhere, whichever way that things like voting may happen to turn out.

Maurice

****

On January 13th, 2021 at 6:53 PM, Bill Walker sent me the following message, with "Are you proud" as the subject:

Are you proud of your vote for Donald Trump? Do you take responsibility for the disgusting assault on our Capitol? Are you still brainwashed or are you able to hear only Rush?

Bill

Sent from my iPhone

****

On January 14th, 2021 at 5:31 PM, I responded to Bill with:

Bill,

The things are a mixed bag. I still continue to consider talk radio, but I also continue to consider the Dalai Lama and other sources of perspectives.

His Holiness The 14th Dalai Lama supports the well-being of all sentient beings and has mentioned before that he supports Marxism, though not supporting Totalitarianism.

I respect both Capitalism and Marxism, and I respect how the pendulum swings back and forth between Conservatism and Liberalism in the U.S., as well as between more free markets and more redistribution. I respect both major political parties, and if the

Democrats' recent increase in power and their current push in the direction of Liberalism results in a major degree of Marxism someday in the United States, then I view that as a mixed bag as well.

Things have an ebb and flow, and I can curtail my speech to a major degree to be cautious of the government and big tech. The First Amendment has major value, yet if the nature of our government goes into directions limiting speech out of concerns for safety, then I can in many cases limit my speech. The 14th Dalai Lama has mentioned before that he considers himself half-Buddhist and half-Marxist. With the recent extra power obtained by the Democrats, and with the push by Bernie Sanders and Alexandria Ocasio-Cortez to take things further in the direction of Socialism, I am prepared for The United States government potentially moving into a Socialist future, even a Marxist future, especially out of my respect for how His Holiness The 14th Dalai Lama has spoken favorably of Marxism. I value how some Democrats and Republicans support of the value of Capitalism, and I value how The 14th Dalai Lama supports Marxism, and I value how some Democrats support Socialism.

I am neither proud nor ashamed of my voting record for presidential elections: 1996 skipped, 2000 Bush in the General, 2004 skipped, 2008 Obama in the General, 2012 Stein in the General, 2016 Cruz in the Republican Primary, 2016 Clinton in the General, 2020 Trump in the Republican Primary, and 2020 Trump in the General. From a Buddhist perspective, I am open to both the possibility of someday voting for Ted Cruz in a general election for U.S. President and the possibility of someday voting for Alexandria Ocasio-Cortez in a general election for U.S. President.

Namo Varjadhaka.

Maurice

＊＊＊＊＊＊＊＊

Bill's response, at 1:57 PM on January 19th, 2021 was:

Your perspective is interesting, and I appreciate the opportunity to have a dialogue with you about this

Bill

＊＊＊＊＊＊＊＊

At that stage, I thought that Bill Walker and I had made it to the other side of the political bickering. For a number of years, he had been one of my professional references, and he had sometimes been a legitimate friend.
　However, something changed about the total effect of the political situation on his heart and mind.

＊＊＊＊＊＊＊＊

On January 21st, 2021 at 8:31 PM, William Crockett Walker sent the following e-mail message to me, with "Think next time" as the subject:

Next time you vote think.

You voted for a man who's supporters stormed the capitol and killed a cop. You voted for a man who separated babies from mothers and locked them in cages. These are facts not opinion

Your vote put my children in danger

Your vote put my country in danger

I know you never served this country but you father did. How do you think your father would fee seeing people wearing Trump hats breaking windows in the capitol and beating policemen?

I was your friend. I even wanted to hire you but your support of Trump has cost you both of those.

Think next time your vote it's important

Sent from my iPhone
................................***....................................
[Note: Yes, he skipped periods several times, had "but you" instead of "but your," and "fee" instead of "feel."]
................................****...........................
To me, he had really crossed several lines. Also, the entire situation in the entire country had crossed many redlines. I decided to hold nothing back in forcefully responding, something I almost never do.

Here is what I responded to that message with, including him on the "to" line and The Republican National Committee on the "cc" line, on January 22nd, 2021 at 11:53 AM Houston time:

Bill,

Next time you vote think.

There is a bigger picture than your presented obsession with some of Trump's comments and the acts of some of his supporters.

There are catastrophic dangers and opportunities presented by both the far right and the far left in this country and around the world. There are also dangers and opportunities presented by the full political spectrum.

There are legitimate reasons why dozens of millions of Americans voted for Biden, and there are legitimate reasons why dozens of millions of Americans voted for Trump.

Talk radio tends to be sharply critical of how some Black Lives Matter and Antifa activists turned violent, yet tends to neglect the bigger picture of how Black Lives Matter and Antifa to a major degree fit into to the process of benefiting America's development.

The mainstream media tends to be sharply critical of Trump and Trump supporters, yet tends to neglect the bigger picture of how Trump and Trump supporters to a major degree fit into the process of benefiting America's development.

The right wing tends to ignore how a degree of holding back on saying "Blue Lives Matter" and "All Lives Matter" is a holding back that illustrates the challenges of how many Black Americans face an extra degree of challenges and burdens. The right wing tends to ignore how a degree of saying "Black Lives Matter" illustrates the challenges of how many Black Americans face an extra degree of challenges and burdens.

The left wing tends to ignore how a degree of holding back on saying "Black Lives Matter" illustrates the challenges of how some White Americans, Asian

Americans, Native Americans, People of Mixed Race, and Others face an extra degree of challenges and burdens brought on by People of All Races and Factors Beyond People. The left wing tends to ignore how a degree of saying "Blue Lives Matter" illustrates how Police Officers of All Races face an extra degree of challenges and burdens presented by Americans of All Races and Factors Beyond Americans. The left wing tends to ignore how a degree of saying "All Lives Matter" illustrates Dr. Martin Luther King, Jr.'s ideal of judging people "by the content of their character" and illustrates the value of some perspectives from the New Testament of *The Holy Bible*, the TV series *Kung Fu*, and other sources.

Many people from both the left wing and the right wing tend to recognize that there is good reason to usually hold back on giving voice to the saying "No Lives Matter," because it has the potential to be a catalyst for human rights violations, genocide, and other such things. However, many people from both the left wing and the right wing are likely oblivious, at least at the conscious level, of how on rare occasions giving voice to the saying "No Lives Matter" helps to illustrate a degree of reality in some perspectives from the Book of *Ecclesiastes* and *The Dhammapada*, in which reality transcends the dichotomy of lives mattering and lives not mattering.

Many supporters of the Military-Industrial Complex tend to ignore the value of how Trump was instrumental to The Abraham Accords, and many supporters of the Military-Industrial Complex undervalue Peace.

Many supporters of Peace undervalue War.

Many supporters of War undervalue Peace.

Many supporters of Election Integrity undervalue The Occult.

Many supporters of The Occult undervalue Election Integrity.

To some degree I support War, Peace, Election Integrity, and The Occult. To some degree I exist. To some degree I do not exist. To some degree both You and I exist. To some degree neither You nor I exist.

The Buddhist orientation of Nonattachment, Nonaversion, and Nonindifference has a degree of applicability to the fabric of reality.

The Christian orientation of Love for All Human Beings has a degree of applicability to the fabric of reality.

The Marxist orientation of Class Warfare and Pathways to Resolve It has a degree of applicability to the fabric of reality.

The Capitalist orientation of Free Market Competition has a degree of applicability to the fabric of reality.

Aleister Crowley has a degree of applicability to The United States of America.

The Ten Commandments from *Exodus* 20:3-17 have a degree of applicability to The United States of America.

The United States Constitution has a degree of relevance to The United States of America.

The Bill of Rights has a degree of relevance to The United States of America.

Namo Vajradhaka.

Maurice

****

The tipping point of my decision to change from the prior plan of seeking to publish this book in early 2024 to seeking to publish this book in mid-2023 was when I learned on June 11th, 2023 that Honduras had recently ended diplomatic relations with Taiwan (Republic of China) and had recently started diplomatic relations with Mainland China (People's Republic of China).

****

Here are a few conversational items from the May-to-June 2023 period:

A man named Brian Player spoke with several people in a group setting about multiple big screen portrayals of Batman and related those to portrayals of Superman and other comic book superheroes. I stayed silent for a very long time, as he spoke of advantages and disadvantages of different portrayals and issued some criticisms about the role of access to wealth and technology as fundamental to his abilities.

Eventually, I said something like, "Wealth itself is not enough for someone to really get stuff done at such a profound level. They still have to have the abilities to coordinate and use the wealth. Also, there's the issue of how, in many of those movies, if you really look at what they portray, the heroes who lack clearly identifiable superpowers beyond the use of high-tech tools still, when fighting gets really difficult, show abilities way beyond what normal people can do. For example, you see someone endure something that almost anyone in the normal range of physics and such would not have been able to endure. Actually, there are reports that ordinary people, historically, in some extreme situations wind up with temporarily achieving super abilities to make things happen. For example, there are reports sometimes that a person discovers a loved one to be trappped under a heavy object that normal physics and situations would lead us to believe there's no way the person could lift, yet the person lifts it anyway as part of saving the life of the loved one."

Brian responded with something like, "Yes, but I don't think anyone here in this room can fly and shoot lasers from their eyes."

I answered very much in the manner of, "That's probably true. But sometimes really weird things do happen. I've experienced some of them firsthand. Although I don't know if Alex has read that part of it yet, the book I gave him one copy of for his birthday includes in its nonfiction epilogue information about when I literally experienced a time loop. I've seen some parts of the movie *Groundhog Day*, but not other parts of it, and what I experienced was very much like that, though for me it was only one loop. *That's the kind of thing that can really get your attention in life; I mean to think, it's something that in regular life seems to defy almost all of the textbook physics, yet I experienced it directly.*"

Although the referenced Alex (who has indicated to me that he is of the family name of Pena) did not seem to chime in with any strong reaction at that time, later on it came up in conversation again.

In June 2023, Alex Pena and I were discussing the

age-old debate about how, when lost-and-found money crosses a legally-stipulated threshold, it then becomes the legal duty of the person finding it to report it to the authorities. I mentioned that I would expect to totally adhere to that standard, with the exception of if I was in an emergency, such as separated from means to reach safety in general and/or necessary help and the use of some of it would help obtain transportation and other necessities. In that scenario, I stated that I would use what was needed to get through the situation ethically and then expect to meet with the legal authorities and explain the necessity of using that portion of the funds.

At several points in that conversation he seemed flabbergasted at my expression of it not mattering whether it was $1,000 or over a trillion dollars, that I would focus on doing what I believed to be ethical, which in the vast majority of cases would include obeying the law.

When he expressed that he did not feel quite such an extreme commitment I mentioned uncertainty about whether he had noticed my mentioning the September 27th, 2019 time loop to Brian a few weeks earlier and uncertainty about whether he had read the epilogue to *The Dimetrodons, the Dorians, and the Modern World: Revised Edition* yet. Also, I tied this together with surreal actual experiences being one of the reasons for my extreme commitment to doing what I believe to be ethical, out of respect for *knowledge* that the deeper reality can *and sometimes does* deliver unto people things that many would consider unthinkable prior to suddenly becoming forced by reality to deal with the profundity of *directly experiencing* the hitherto-unthinkable.

Instead of expressing straightforward skepticism together with critical thinking, he resorted to mocking me by saying, "You must have been dreaming."

I explained, *"It was not dreaming. It was waking life. I could tell. It was as real as you and me right here and now."*

At some stage, he disparagingly joked, "Oh, I 'know' what it was. It was daylight savings."

I mentioned something like, "Look, you can believe or not believe as you choose, but I *know* what I experienced with that."

Later on, he changed tactics, saying with playful sarcasm, "Now I can tell people that I have a coworker who is a time traveller."

I responded, in essence and nearly verbatim, "It would probably be better for you to say that you have a coworker who alleges that he is a time traveller. From where you stand you do not know for sure about it, though *I know from having directly experienced it.*

"This way you would have better purity of speech. That being said, it was only that one time, as far as I know, and *I don't know how or why it happened."*

Alex, Brian, and I are coworkers at Goodwill Houston. At the time of publication of this book, I have three jobs: a sole proprietorship self-employment position that published the three fiction novels that led up to this nonfiction book, CEO/Founder of the C corporation that is the publisher of this very book, and a basic-level position as an employee at Goodwill (which at present complements the othe two vey well). Meanwhile, three of my supervisors at that nonprofit organization are the aforementioned Brian, the aforementioned Alex, and a woman named Crystal Brown.

Crystal and I had weeks earlier shared an interesting conversation in which she inquired about daring past behaviors, and I revealed specifics about a few extremely adventurous choices that led to ultra-bizarre experiences other than the aforementioned timeline discontinuity event. When I returned the favor by inquiring about some of her past daring behaviors, she revealed a similarly bizarre experience: an incident that she can use as testimony about believing herself to have profoundly experienced God.

A little later, there was a follow-up conversation, in which I found out that, per her account, when she witnessed Jesus Christ in sensory perception she *knew* somehow it was Jesus Christ and not an impostor or a coincidentally-similar being. I mentioned that it seemed to me that this is similar to how sometimes and in some ways beyond the five senses there is *often* a kind of *knowing* of the *presence of mind* of another being, especially if there is a familiarity with that being. In contrast, with my strange experiences I have often perceived major uncertainty about the degree(s) to which God &/or aliens &/or alternatives were at work.

On June 28th, 2023, Crystal and I were in the break room sharing an additional conversation, which went something like this:

Maurice: Have you finished watching *Raiders of the Lost Ark* yet?

Crystal: No. I was on vacation recently.

Maurice: You *are* taking a while to finish viewing it. That being said, I remember that from something like February to early August 2021 I spread out the viewing of a movie that is in some ways very similar to *Raiders of the Lost Ark* and in some ways very opposite to that movie.

I completed a viewing from February to something like August 5th or August 4th, 2021 of *The Song Remains the Same*.

Crystal: That's neat!

Part of the background of this was that Crystal had mentioned several weeks earlier to me that she had started her first-ever viewing of the aforementioned 1981 Indiana Jones movie, via an online streaming service.

*****

Next, here are copies of some messages to others, together with explanatory footnotes:

Date: Mon, 09 May 2022 08:22:32 PM CDT
From: Maurice J. Blair
To: Katherine Navarrete
Subject: 2022.05.09 follow-up to 2011 list idea from some emails

Kathy,

On October 1st, 2011 I wrote to you,

""Some time when I set aside enough minutes, I plan to watch "Telephone," Enrique Iglesias' "I Like It," Fat Boy Slim's "Weapon of Choice," Beyonce's "Sweet Dreams," and Madonna's "Bad Girl," under the plan we discussed earlier this year.""

Yesterday at an event on a farm with an address in Cleveland, Texas, a recent Rice University graduate performed karaoke of several Taylor Swift songs and a few songs by other artists. That Rice graduate and I interacted in a reasonably favorable way at the event, though there is significant uncertainty (from my perspective) as I write this email message to you about whether she and I will meet again or share other direct communications (after the Facebook friend request that I sent her last night).

One of the songs that she (Lyla) did not perform at that event was the Taylor Swift song "Blank Space." She and I also did not directly speak of that song's existence in our conversations. However, today, about a quarter-'til-seven in the evening in Houston, I decided to watch that video. That was after having enjoyed listening to the song many times over a number of years without ever having seen the video. After watching that video (an intriguing and inspiring first view, by the way), I remembered that many years earlier I had mentioned to you a plan to watch the above-referenced Madonna video, and that there were other videos in a list with it.

After that, I proceeded to watch the music videos in the aforementioned 2011 list.

Interesting videos, and something peculiar I thought of early in the "Weapon of Choice" video was how Christopher Walken is in both that video and the "Bad Girl" video. Then I thought of how sometime around 1987-1989 I watched a major percentage of the film Dead Zone on television, and how I found it interesting and peculiar. I wonder what your husband Christophe thinks about Christopher Walken.

On a related note, I wonder what (the) Katherine Howard (who was for a while married to Henry VIII) would have thought about "Blank Space" and what Henry VIII would have thought about Lady Gaga's "Telephone."

Maurice

****

Footnotes to that message:

1. A more complete version of the name of Kathy Navarette's husband is Christophe Maso. He wrote a brief book titled *Scream of the Butterfly*, but I have only read a few pages of it thus far. Although Caroline Myss described early in *Anatomy of the Spirit* that some traditions and experiences point toward people having a degree of a soul deposit with any incomplete project, a more complete view could include both that idea and its inversion (i.e., having a degree of soul deposit with any completed project).

From a more all-encompassing perspective, one could bring together the expressions of: 1) soul deposits for incomplete projects, 2) freedom from soul deposits for completed projects, 3) freedom from soul deposits for incomplete projects, 4) soul deposits for completed projects, and 5) hybrids of two-or-more of the previous four. While there is much to be said for the value of completing projects, there is also much to be said for the artistry of coordinating and balancing multiple completions, incompletions, progressions, &c.

2. On a related note, over a span of about 12 years up to the time of publishing this book, I have only read about 30% of *Anatomy of the Spirit*. Another book I have read and carefully considered fractions of over a long span of time is *Are Men Necessary?: When Sexes Collide* by Maureen Dowd.

I bought *Anatomy of the Spirit* and followed up by a pattern of occasionally reading some of it, setting it aside for a few days or a few years, then reading more. in contrast, I checked out *Are Men Necessary?* from a library in mid-2011, read something like 1/4 of it, returned it to the library, then did not get back to reading it until about 10:30 PM on April 30th, 2023, after receiving a copy that I had recently purchased from an online vendor. I went back to restarting from the beginning and have made limited progress with it thus far.

In contrast with both of those books, I completed initial beginning-to-end readings of the six books and one short story of the bullet list on page 141 long ago (having finished first complete readings of each of the books multiple years earlier than my September 2021 initial complete reading of the short story).

3. The October 2011 list of music videos emerged from e-mail correspondence in which Kathy and I would inquire with each other about popular songs and music videos. Both sides would issue recommendations.

* * *

Sometime soon after Jeff Hancock posted on his Facebook timeline on August 19th, 2022 a link to a story about controversies regarding the aftermath of the U.S. Supreme Court's *Dobbs v. Jackson Women's Health Organization* ruling, I posted as a response:

For many recent years, the debate about "pro-life vs. pro-choice" is something I decline to land exactly anywhere on in terms of policy stance. When I was younger there were some times I landed on one side and some times I landed on the other. One of the odd conversations at Duke, though, circa March 1996, was when Jennie Lee and Heather Zorn were discussing with several other students this subject, and I started at some stage saying something very much like, "I believe that even in cases not involving things like rape and incest" and they interrupted me to preemptively steer things away from having me finish the sentence. I decided to decline to tell them or anyone else at Duke what I had intended to finish that with... until now. At that time, I was in a state of heightened infatuation with some women, especially Lindsay Schneider (whom I dated and quasi-dated a few times in the February to May 1996 time-frame), and I intended to complete the sentence with something like, "... abortion should generally be legal." In contrast, much of the time mid-2000 to circa March 2003 I was very much in favor of ideas like, "... abortion should generally be illegal." A key example was that at a Halloween 2001 party in Austin - a very rowdy party, mind you - one of the many people I met was a woman I might have otherwise pursued, but in her basic introductions she mentioned that she worked for Planned Parenthood. At that time, largely influenced by a multitude of factors, I felt major discomfort with the idea of pursuing dating an employee of that organization. Two related twists to this are that the aforementioned Lindsay had mentioned that she was very involved with a Young Republicans group in high school or a similar stage, and that on 6/21/2009 and 5/22/2010 I sent FB messages to her (though here surname had an update). The first of those was a normal-style person-to-person message. The second was simply "Namo Vajradhaka" , and she almost immediately blocked me on Facebook as a response. In debates of preservation vs. destruction vs. creation &ctra, it is generally good - yea verily, very good - to have a reasonably healthy respect for the The First Amendment to the U.S. Constitution.

****

Footnotes to that message:
1. A few times I visited Lindsay Schneider's dorm room, and she was kind enough to lend me some Madonna music, which I returned to her in a spirit of goodwill.

2. Lindsay accepted my invitation for her to go on a movie date with me to watch *Casino* in the movie theater located on-campus back then. It was in or near February 1996 when we attended that Scorsese-directed film.

3. Her name would eventually display on Facebook as being Lindsay Schneider Morris. The "here surname had an update" phrase could translate into "on this here Facebook platform, that surname had an update."

* * * * * * * * *

Next, we delve into how for many years a Tarot Article written by yours truly was publicly-available on the Internet, but eventually disappeared from the Internet, and we revisit the unedited first-submitted proposal for that article.

Initially, on October 13th, 2008, as measured by Houston, TX time, I submitted an article idea to the web page Aeclectic Tarot (http://aeclectic.net/). Here are some of the fields of that submission:

article's name: Perhaps The Dawning of a New Tarot Deck

Has this article been published before? If so, where? No.

Article Submission:

Part One: The Main Framework

Here is a possible framework for a new Tarot deck geared strongly towards big business:

Four suits of the minor arcana:
* Marketing
* Operations & Management
* Public Relations
* Finance & Accounting

The fourteen cards within each suit of the minor arcana:
* 1 through 10
* Staff
* Supervisor
* Vice President
* President

The twenty-two cards of the major arcana:
0. Conflict
I. Chemistry
II. Entrepreneurship
III. Chief Financial Officer
IV. Chief Executive Officer
V. Business Philosophy
VI. Salesmanship
VII. Competition
VIII. Honesty
IX. Reflection
X. Solvency
XI. Motivation
XII. Sacrifice
XIII. Bankruptcy
XIV. Action
XV. Deceit
XVI. Crisis
XVII. Goals
XVIII. Uncertainty
XIX. Confidence
XX. Profitability

XXI. The Economy

Part Two: Background of Its Development

In 2001 to mid-2003 the author of this article studied a number of esoteric subjects, but knew little of Tarot. In mid-2003, the author started studying the Tarot deck, and over the years that followed, purchased a copy of The Oswald Wirth deck and a copy of The Rider-Waite deck. One of the influential writings on Tarot for the author is P.D. Ouspensky's 1913 essay "The Symbolism of the Tarot," which is currently in the public domain.

On September 27th, 2008, the author contemplated business, Tarot, and related subjects, then composed the following framework for a possible Tarot deck:

Four suits of the minor arcana:
* Marketing
* Finance
* Management
* Accounting

The fourteen cards within each suit of the minor arcana:
* 1 through 10
* Staff
* Supervisor
* Vice President
* President

The twenty-two cards of the major arcana:
0. Conflict
I. Chemistry
II. Entrepreneurship

III. Chief Financial Officer
IV. Chief Executive Officer
V. Business Philosophy
VI. Salesmanship
VII. Competition
VIII. Honesty
IX. Reflection
X. Solvency
XI. Motivation
XII. Sacrifice
XIII. Bankruptcy
XIV. Action
XV. Deceit
XVI. Crisis
XVII. Goals
XVIII. Uncertainty
XIX. Confidence
XX. Profitability
XXI. The Economy

In the days that followed, the author contemplated a possible Rider-Waite-style inversion of Card VIII And Card XI, together with an added twist of altering Card XV. This variation would have the major arcana feature:

VIII. Motivation
XI. Authenticity
XV. Obstacles

On October 7th, 2008, the author considered varying the September 27th version by replacing the "Management" suit with an "Operations" suit.

Then on October 12th, 2008, the author decided on altering the suits further, leading to the grouping described at the beginning of this article ("Marketing,"

"Operations & Management," "Public Relations," and "Finance & Accounting").

Also, on October 12th, a search of the web did not reveal any existent Tarot decks with anything closely resembling the framework described here. However, with the large number of people involved with conceiving Tarot decks, it does not guarantee that there is no one else who came up with something somewhat similar as of that date.

Part Three: Main Alternate Version

This means that the main alternate version to the main framework mentioned earlier would be:

Four suits of the minor arcana:
* Marketing
* Operations & Management
* Public Relations
* Finance & Accounting

The fourteen cards within each suit of the minor arcana:
* 1 through 10
* Staff
* Supervisor
* Vice President
* President

The twenty-two cards of the major arcana:
0. Conflict
I. Chemistry
II. Entrepreneurship
III. Chief Financial Officer
IV. Chief Executive Officer

Part Four: Second Alternate Version

Because P.D. Ouspenksy's 1913 essay "The Symbolism of the Tarot" included both the Rider-Waite VIII/XI inversion and an additional inversion of V & VII, there can be another alternate version of the deck proposed in this article: the same as the main alternate version just outlined, except with "V. Competition" and "VII. Business Philosophy."

Part Five: Possibilities of What May Be Next

If there are artists interested in exploring creating artwork for decks with any one or more of the frameworks described here, the author of this article may be interested in a joint venture pursuing major publication of the deck or decks that result.

Note: The author composed this article on October 12th, 2008.

***

Footnotes:
1. Solandia, the webmaster of that webpage, requested that I expand the article. I complied with that, and the article eventually became published on June 16th, 2010, including some additional revision and editing after the expansion.
2. Over a number of years, I would often skip visiting the article for several months, then check to see if it was still there. They evidently removed it sometime within circa the December 2020 to March 2021 period.

****

Please bear in mind that by sometime 2008 my father and I had both viewed all four of the then-existing *Indiana Jones* motion picture films (with first viewings of the first three in the 20th century and with first viewing of the fourth during its main theatrical run). I plan to watch the fifth motion picture film in that franchise sometime after making sure that this book reaches publication.

****

Next is a composite of portions of three electronic messages (one each from May 25th, 2022; June 3rd, 2022; and April 8th, 2023) together with editing:

Perhaps the funniest thing I've experienced recently was when people were debating reality on Facebook on multiple pages and I noticed some repeating an idea that has cropped up for hundreds and perhaps tens of thousands of years: an idea of relating religiosity with childhood behaviors of emphasizing imaginary friends, and choosing to keep myself limited to skipping commentary on some debates and making seriously elliptical references as commentary on some debates. You are probably aware that in mid-1996 to October 1996 my mind went haywire after I had attempted to simulate lysergic acid diethylamide (LSD-25) via heavy dosages of listening to 1965-1967 music and experimenting with directing and redirecting my mind in many directions. Part of the reason for the later shifts was involved with healing from that and other challenges in life. To the best of my knowledge I have never in this lifetime consumed LSD-25, yet I am thankful for how in the early-to-mid 1990s I yearned for paranormal experiences and eventually I had a mixture of some experiences that may or may not have been paranormal... and some experiences that were - from my vantage - incontrovertibly paranormal.

Perhaps the most incontrovertibly paranormal experience I had was in Houston during part of the morning of July 10th, 2005. Evidently, the address was 7601 Fannin Street. Previously, something also incontrovertibly extraordinary though in some ways more threatening had happened on July 8th, 2005

Here is a timeline:

1998: Stable treatment and life activities; however, in mid-1998 I embraced a bleak outlook on many areas of reality.

1999: Bleak outlooks continued, though I opened up to being more agnostic toward many things.

2000: With cooperation from Dr. Jenna Saul, I ventured into stopping medication, though some *Twilight Zone*-type stuff happened and I soon went back onto medication. The emergency room did not require me to have in-patient treatment, though.

2003: At an extremely low dosage of one psychiatric medication things were going well for a while, then some weird things happened and I ended up in Harris County Psychiatric Center for something like three-and-a-half weeks straight. I came out with a heavy set of multiple medications.

2005: With cooperation from Dr. Sateesh (of MHMRA), I tapered down to being completely off of medications. Then a project I had started years earlier came to fruition, in which I composed and/or channeled an attempt at a magnum opus of a novel (semi-anonymous or anonymous, though in some conventional senses written by me). At 73,000+ words, I registered a copyright with it as an officially anonymous work, with myself as the claimant, and declining to register an ISBN. It was at many levels a deliberate and willful attempt to meddle with all of the most dangerous areas of reality in the hopes that whether I would live or die soon thereafter, extraordinary levels of awareness would be liberated. Some of the bizarre things that happened bordered on the incomprehensible with respect to normal life. I ended up going to an emergency room on July 8th, 2005, then getting transferred to Intercare/Intracare (evidently 7601 Fannin St). Part of what I experienced on July 10th, 2005 was something that about seven years later I recognized to have major similarities to part of what likely happened to the hikers in the Dyatlov Pass Incident, though the official medical descriptions (of what happened to me) were likely a very whitewashed simplification and distortion of what truly transpired. I never saw those records, by the way. Part of their resolution to when an unknown and compelling force evidently pinned my brain at a fixed altitude while I was seated on the floor in a left-foot-atop-right-knee cross-legged position was that, after a while, they told me to uncross the legs. Then, about four men lifted me up and placed me face-down on my bed. People then injected me with a cocktail of two very different mind-altering prescription drugs. Very soon after that, my right kneecap felt like it was going to explode. To counter this, I used a combination of religious languages and concepts, a method I had preplanned for some bizarre contingencies: specifically, to shout, "Yang... Yin Yin Yin Yin. Yang... Yin Yin Yin Yin" repeatedly. Suddenly, my kneecap no longer felt that extreme pressure. There things were, something like 10:45 AM that day, to the best of my memory. Then I stood up to walk away. Then everything vanished. I was suddenly in an altered state of mind in which -- as far as I remember -- words and concepts were completely absent from my mindstream, though I was awake and/or in a mixture of wakefulness and vivid dreaming. Spatiotemporal relations and the lighting seemed very altered. I witnessed myself seated, then choosing to stand up and turn, with my parents present, then felt my head bump into something, ending up knocked out cold. My father later revealed to me that in his experience, he witnessed me walking from my room toward the area outside the room (where he was) and seeing me walk right into the doorframe on the left side of my face. I woke up on July 11th, 2005.

2018: I had been successfully stable on a very reduced daily dosage of medication from something like the later part of 2009 onward for years, and was consistently visiting Dr. Gerald Busch only once a year. I especially looked forward to visiting with him in late 2018, because he set terms by which I could drop from the only medication I was on from half of the normal minimum dosage to one-quarter of the normal minimum dosage. If I would be successful enough with that dosage over a reasonable length of time, then he could grant to me another attempt to stop taking that drug, with doctor approval. However, I found out that he was no longer even practicing psychiatry. I felt at that time that Dr. Busch was the only psychiatrist I could wholeheartedly trust with the peculiarities of my situation, and the decisions I made included skipping treatment from the PACE clinic that had bought out his practice.

2019-2021: Additional unusual things happened. However, fortunately, I have generally been doing well, both mentally and physically, since partway into the day on July 12th, 2021, and I had my last dosage of a psychotropic medication on May 28th, 2021.

If things continue going reasonably well, I may continue being a former patient of the psychiatric industry rather than being a patient of that industry.

Some of the behavior, judgments, and mental distortions exhibited by psychiatrists and other healthcare workers have been problematic for many people over the centuries, yet I believe there could be major opportunities for them to do better.

****

Footnote: The July 8th, 2005 incident involved multiple tense and bizarre dynamics culminating in, for an extended amount of time, finding my automatic central nervous system processes completely losing the natural and automatic inhalation process though retaining the automatic exhalation process, accompanied with a sudden involuntary earworm of the last few seconds of the "Suicide is Painless" theme song often associated with M*A*S*H on repeat. That led to my agreeing with my father that I should visit an emergency room, yet at the emergency room I found myself unable to directly tell them about that condition, instead focusing on a July 7th, 2005 incident in which I heard a Houston radio advertisement description of a scenario of alien abduction followed by extraterrestrial aliens vivisecting a human being, then segueing into either reminding the listener that he or she could save money on auto insurance or something similar. I made a semi-coherent reference to how when I heard that advertisement, I attempted to do an improvised chöd ceremony simultaneous with the description of the abduction and vivisection, placing myself in the plane of the imagination as matching the story character's situation of getting trapped and experiencing involuntary vivisection. Chöd is one of the practices of Vajrayana skillful means, and it can involve many different scenarios of imagining physical danger and harm to befall oneself, yet it normally does not result in actually all of a sudden experiencing a clear and distinct tactile sensation of the imagined scenario happening to the body. In my case, though, I suddenly felt physical sensations that seemed identical to what it would be like for aliens to have my brain exposed to scalpels or similar cutting instruments without anesthesia and then for the surgeons to perform extremely rapid brain surgery as part of digging for something.

I felt a presence of mind(s) as this happened, then I felt that presence identify the Churchland description of Eliminative Materialism and adjacent prior lodging in my brain of a structure expressing a page not previously witnessed directly by my conscious mind on any page and previousy recorded, but a subconscious item adjacent to much of that Paul Churchland and associated thinkers' written expressions, a page bearing the writing, "You have no mind." All of a sudden, the tactile sensations of what felt like brain dissection while alive and fully awake ceased, and I felt an intangible impression of telepathic awareness of the presence(s) responsible fleeing in horror in response to seeing that pronouncement. Of course, if we consider things like Jesus Christ reportedly saying, "I have come to blind those who see and to bring sight to those who are blind" (Cf. *John* 9:39) and Arthur C. Clarke stating, "For every expert there is an equal and opposite expert" (reminder of reference: Clarke's Fourth Law), then we can recognize that there is what some might call a spooky remote relationship between the idea structure "You have no mind" and the idea structure "You have a mind."

I spoke only briefly about part of that, and somehow did not find it within myself a way to bring out into the open for the healthcare workers the fact that I kept having to use volition to consciously override the modula oblongata and other things just to make each new inhalation happen with which to stay alive. Also, there had been the spectre of how an excerpt of the concluding instrumental flourish of a performance of the aforementioned Johnny Mandell composition accompanied how right after I had said out loud, "I resign my will" the sudden-onset loss of automatic inhalation started; those combined with other factors had suggested to my conscious mind that I might have crossed one line too many in relationship with God/Reality/Whatever/Whoever and was in a state between having unintentionally performed an act fatally causing the second death (i.e., annihilationist death; what people in some areas of theology believe to be when a person reaches such a final state of apostacy that he or she becomes totally, permanently, and irrevocably obliterated by God, dying in terms of consciousness to an absolute degree such that some might consider it to be a permanent wiping out, almost equivalent from that being's own perspective of consciousness to having never been born, never lived, and never existed; in short, a total extinction of consciousness with no hope of ever returning to consciousness; what some might term absolute termination). Perceiving myself at that time to have been one missed system override breath away from such a fate, this at some levels struck me as an appropriate comeuppance unless I could make perfect-enough choices to recover from the predicament, which seemed especially appropriate to find myself in after having explicitly explored much of the mystery of real or alleged absolute obliteration and many other mysteries in portions of *All Things Under and Over the Sun and Stars: An Enigma in Twenty-Three Stages* {completed from mid-2001 to June 3rd, 2005; published on a nonprofit basis on June 4th, 2005 with extremely limited distribution and an officially anonymous attribution of authorship; performed through a mixture of a) attempting to channel all versions of The Absolute/Almighty/Creator/Destroyer/Ultimate/Totality, all other celestials, various deceased authors (such as J.R.R. Tolkien, John of Patmos, and Nostradamus), and other beings, b) attempting to remote view across mutiple real and actual universes within the multi-verse or across multiple real universes across multiple multi-verses and to then report on the findings within the guise of a fiction novel that would double as a portal, c) relating some semiautobiographical details in the guise of a few choice portions here and there, d) more traditional novel-writing methods, and e) more; later serving as much of the basis for what would become the novel *All Things under and over the Sun and Stars: Enigmas in Various Stages* (published in two differently-sized hardcover formats on January 2nd, 2023 and a third hardcover format and a paperback format on January 3rd, 2023; each including the potential for profits as one of the declared motives), which attributes to myself its official authorship}.

One of the reasons for the difference in attribution of authorship was that I had felt with the 2005 work that jumping right into attributing such to myself would be tantamount to a distortion in the direction of an extreme God complex; whereas I truly believed in early 2023 and still truly believe at the time of completing this follow-up work in mid-2023 that attributing the January 2nd/3rd, 2023 work to myself is not tantamount to a distortion in the direction of an extreme God complex, given changes in the times, seasons, and purposes of the realities.

By the way, it seems that multiplying the phrase "Eliminative Materialism" by two levels of *John* 9:39, Clarke's 4th Law, or something else that can function to fully reverse idea structures, coud result in the phrase "Reintroductive Idealism," the phrase "Retentive Immaterialism," and/or the phrase "Conservational Spiritualism."

By the way, circa February 2006, I said out loud, "I accept my will" (to counterbalance the prior statement.)

* * * * * * * * *

One of the dividing lines in the sands of society involves conceptualization of what constitutes mental illness, what constitutes sanity (i.e., sane mental health), and whether there exists the ability to grow out of mental illness(es).

There are people who believe that once a person has been officially diagnosed by at least one licensed clinical psychiatrist as having a major mental illness, the person should be societally and cognitively branded for life as mentally ill and/or much more susceptible than the average human to becoming mentally ill.

In contrast with that, there are people in a variety of areas of many societies, including those using some interpretations and methods described by Buddhism, Christian Science, and Scientology, who believe that people are capable of learning, growing, and healing such that once a person has been officially diagnosed by at least one licensed clinical psychiatrist as having a major mental illness, the person should not be societally and cognitively branded for life as mentally ill and should not be societally and cognitively branded for life as much more susceptible than the average human to becoming mentally ill.

Although I recognize many levels and degrees to which reality transcends idea structures and narratives, if given a chance to cast a ballot at the time of composing this book for which of the two camps just mentioned should be more applicable to how I currently relate to psychological resiliency and mental health, then I would vote for the view that refrains from branding people for life in those ways.

Of course, some of those who have never been diagnosed with serious mental illness might point out that this undercuts much of their feeling of security and self-esteem in terms of taking pride in having always avoided such diagnoses and in looking down on those who have received such diagnoses, yet, as boxing instructor Jesus Poll was fond of telling me at Savannah Boxing Gym in a few portions of the early 21st century when I would finally start getting a technique right, Exacto!!

Consider if someone were to ask me right now without specifying minutia of which way(s) to define possession, "Do you have any psychiatric diagnoses?" If I were to give a simple yes-or-no answer without providing clarification, then, given the fact that I have worked hard for decades to be living as well as I do while successfully being off all psychiatric medications for a long and sustained period, together with how I believe strongly in the ability to grow out of problems and heal, I believe that the most honest answer would be, "No."

Although I have a degree of respect for the challenges, compassion, and good intentions of many psychiatric healthcare workers, I believe that there is a high degree of truth in much of the documentary *Psychiatry: An Industry of Death* (2006), which I first viewed and reflected carefully upon in the April 6th-7th, 2023 period.

What about something where I land on the "societally more respectable side of the fence' as another manifestation of the issue of people showing fixated condescension toward others versus refraining from showing fixated condescension toward others?

I have never been criminally convicted, although a) I did receive a speeding ticket in the second half of 2004 for driving 83 mph in a 65 mph zone, b) I have received warnings from law enforcement authorities for driving significantly over the speed limit on other occasions, and c) I admit that I went skinny dipping late at night at Surfside Beach with several other members of a large group of revelers on a July Saturday night in 2009.

~~~

Also, I admit that near the end of April 2006, a woman who said that she had just returned from being jilted at the altar by a man in Conroe introduced herself to several other people, including myself, at a Greyhound station, spoke in a very ribald way indicating that she was in heat, and proceeded to buy what was presumably an illegal substance for smoking. Those of us at that gathering outdoors near that station, the infamous one close to Minute Maid Park, started passing the peacepipe-style thing around. I had never smoked before, yet when it arrived to me, I decided, *you know what, this amorous lady in my presence might be the one, I will go ahead and smoke the peacepipe kind of item, even though by context it is almost undoubtedly an illegal drug.*

Yes, by the way, in contrast with President Bill Clinton's reference to not inhaling marijuana, I *did* inhale whatever substance(s) were in the thing that the sexually-aggressive female acquaintance had purchased for something like ten dollars from a male dealer there. It felt almost immediately like a partition that my mind and my brain would normally keep in place had temporarily dissolved. I hung out with the lady and the other people gathered there, then the lady, seemingly in tandem with one of the other gentlemen present, suggested that multiple men, including myself, accompany the woman in walking toward a specific parked motor vehicle.

When others were getting into that vehicle I could tell that their plan was for that one very-aroused woman and at least two very-aroused men to get into it at the same time and to have fun and include plenty continued smoking of drugs. I suddenly sensed that for my state of being at that time it would be going too far off the rails of risk management and spiritual decency to get into the vehicle with them.

I suddenly announced to them, either verbatim or nearly verbatim, "I have decided not to get into the vehicle with you here and now, but I wish you the best with your lives. God bless," then walked away. I felt no regret toward the situation. I have not smoked since then, though I still leave the option open to someday return to smoking.

When doctors' offices (and other healthcare organizations) require me to fill out paperwork regarding my medical history, sometimes they have a basic binary option question of "Have you ever been a smoker?" Usually, since either a straight "yes" or a straight "no" would tend to be somewhat deceptive in my situation for how to answer that, I have tended to, when possible, ask someone there about how best to collapse the uncertainty of the balance of factors into either a straight yes or a straight no. Invariably thus far, the answer I have received each time has been to simply answer no.

Since none of the people at the gathering in which the group passed around the mystery substance (about which by indirect evidence, research, and direct conversations, I still have not become very certain about the identity) and took turns smoking it, I remember clearly that each person appeared to very clearly be a full-fledged adult, meaning it appears clear that the statute of limitations in the jurisdiction expired long before the publication of this book.

~~~

With the above stream-of-consciousness adventure complete, back now to the issue of how I feel about being on the "generally more socially respectable" side of people with prior criminal convictions versus people without prior criminal convictions (i.e., being a person without any prior criminal convictions).

Although I can recognize that there is a degree of achievement in being on that side of that divide, I am honest with myself and you about how this does in no way, shape, or form make me *inherently* superior *at all* to any given convicted felon or other convict. Indeed, wherever you might happen to fall within the spectrum of this legal compliance issue and the related spectrum of this criminal-conviction-or-lack-thereof issue (which can fuse from two issues into one easily and which can fission from one issue into two easily - linking vs. delinking criminal activity and conviction of criminal activity), please recognize that you could fill in the blank with a list of any three movies of your choice on this theme, and they would (according to some ways of experiencing them) point toward how the mere fact of having a criminal conviction versus lacking a criminal conviction can often have little or no correlation with the current degree(s) of ethical behavior(s) exhibited by a given being or by a given organization of beings.

CHAPTER TWELVE:
OUTLINE FOR A POSSIBLE URANIUM-235
TAROT DECK, REVISITED

MAJOR ARCANA (consisting of 22 cards)

I. Unity XIX. Execution
V. Religion XX. Moment of Truth
0. Nonduality &/or Multiplicity
XXI. The Great Ultimate
II. Duality III. Methodology XVIII. Unknown
IV. That of All Methods and No Methods

XVII. Excavation	VI. Passion
XVI. Law	VII. Pursuit
XV. Super Chaos	VIII. Super Symmetry
IX. #	XIV. ~
X. Beginnings	XIII. Endings
XII. Ancient Wisdom	XI. Modern Wisdom

MINOR ARCANA [consisting of 56 cards]: Four Suits (Creation, Destruction, Preservation, and Commerce) with 14 cards each: Ace, 2, 3, 4, 5, 6, 7, 8, 9, 10, Servant, Manager, Master, and Conquerer

TRANSITIONAL ARCANA (consisting of 14 cards)

B. Beyond	N. North	U. Universal
W. West	C. Center	E. Fast
D. Dharma	S. South	H. Heartfelt
Voidness Models		Tetragram Models
!	?	!!

CHAPTER THIRTEEN: A DEUTERIUM TAROT DECK
POSSIBILITY, REVISITED

ZA	MGM	GMG
DD	FGM	MGF
ABC	ZYX	EE
ZX	MMM	HHH
YYY	BBB	AZ
	XZ	
	DDD	
	EEE	
	LLL	
	WWW	
XX		XY
YX	CCC	GGG

CHAPTER FOURTEEN: A BUSINESS TAROT DECK
POSSIBILITY, REVISITED

I. Chemistry

II. Entrepreneurship

XI. Motivation

III. Chief Financial Officer

IV. Chief Executive Officer

V. Business Philosophy

VI. Salesmanship

VII. Competition

VIII. Truth

IX. Reflection

X. Solvency

0. Conflict

XXI. The Economy

XII. Sacrifice

XIX. Confidence

XVII. Goals

XVI. Crisis

XVIII. Uncertainty

XX. Profitability

XV. Deception

XIV. Action

XIII. Bankruptcy

Ace, Deuce, 3, 4, 5, 6, 7, 8, 9, 10, Worker, Supervisor, Vice President, and President of:

- Public Relations
- Marketing
- Operations & Management
- Finance & Accounting

XXII. Pendulum of War and Peace

XXIII. Harmony

XXVI. Covenants and Removers

XXVII. Effectiveness

XXIV. Authenticity

XXV. Movers

MAJOR ARCANA (consisting of 22 cards)

I. Unity

V. Religion   XX. Moment of Truth

0. Nonduality &/or Multiplicity

XXI. The Great Ultimate

II. Duality   III. Methodology   XVIII. Unknown

XIX. That of All Methods and No Methods

IV. Execution

XVII. Excavation	VI. Passion
XVI. Law	VII. Pursuit
XV. Super Chaos	VIII. Super Symmetry
IX. ~	XIV. #
X. Beginnings	XIII. Endings
XII. Ancient Wisdom	XI. Modern Wisdom

MINOR ARCANA [consisting of 56 cards]: Four Suits (Creation, Destruction, Preservation, and Revelation) with 14 cards each: Ace, 2, 3, 4, 5, 6, 7, 8, 9, 10, Servant, Manager, Master, and Conquerer

TRANSITIONAL ARCANA (consisting of 16 cards)

B. Beyond	N. North	U. Universal
W. West	C. Center	E. East
D. Dharma	S. South	H. Heartfelt
Voidness Models	?!	Tetragram Models
!!	?	!
	!?	

TTT	DTS	STD
LWL	WLW	YA
CBZ	SJ	DA
AD	JS	BC
CBA	XYZYX	EVE
VEV	MADA	ADAM
DSTP	RRSS	SSRR
PTSD	PTS	STP

Ch. 17: An Alternate Portrait of The Middle of 2023

Composed from Middle Portions of 2011 C.E.
to 12:09 AM & 12:10 AM U.S. Central Daylight time
June 16th, 2023 C.E.

I. Unity   II. Duality
0. Nonduality &/or Plurality V. Religion XX. Moment of Truth
XXI. The Great Ultimate   III. Methodology
VI. Passion   IV. That of All Techniques and No Techniques
XIX. Execution XVII. Excavation IX. ~   XIV. #
IX. #   XIV. ~   IX. #   XIV. ~
VI. Compassion   IV. Execution
XIX. That of All Techniques and No Techniques   XIIII. ~
Tetragram Models   N. North   Voidness Models
W. West   C. Center   E. East
                      S. South
                      B. Beyond   Ace of Destruction
                      U. Universal
                      D. Dharma
                      H. Heartfelt
                      Ace of Creation
        Ace of Preservation   Ace of Commerce
                      XVIII. Unknown

C.E. is A.D. vs. C.E. is not A.D.
C.E. is not A.D. vs. C.E. is A.D.
S.R. is SR vs. S.R. is not SR.
SR is S.R. vs. SR is not S.R.
S.R. is not SR vs. S.R. is SR.
SR is not S.R. vs. SR is S.R.

IS versus IS NOT versus TRANSCENDENCE.

PART TWO:
SELECT FOUR-DIMENSIONAL MODELS OF COORDINATING CONSCIOUSNESSES, PREMISES, ENERGIES, AND RATINGS

CHAPTER EIGHTEEN: SELECT ARCHETYPAL MINDSETS PROFILED SOMEWHAT INDIRECTLY BY COORDINATING THEM WITH ARCHETYPAL MOTION PICTURE PREMISE PROFILES, ARCHETYPAL INTERPERSONAL ENERGY RELATION PROFILES, AND ARCHETYPAL RATING SYSTEMS

I. Archetypal Mindsets:

- Two Versus The Rest #1: +A+B-C-D-E-F-G
- One vs. All Else #1: +A-B-C-D-E-F-G
- Two Versus The Rest #2: +A-B-C-D-E-F+G
- Two Versus The Rest #3: -A+B+C-D-E-F-G
- One vs. All Else #2: -A+B-C-D-E-F-G
- Two Versus The Rest #4: -A+B-C-D-E-F+G
- Two Versus The Rest #5: -A+B-C+D-E-F-G
- Basic Relativism: =A=B=C=D=E=F=G
- Emphasis on One #1: =A=B=C=D=E=F+G
- Emphasis on Two #1: =A=B+C=D=E=F+G
- One vs. Four: -A-B-C-D=E=F+G
- Emphasis on Two #2: =A=B=C=D+E+F=G
- Emphasis on One #2: +A=B=C=D=E=F=G
- One vs. All Else #3: -A-B-C+D-E-F-G
- One vs. All Else #4: -A-B-C-D+E-F-G
- One vs. All Else #5: -A-B-C-D-E+F-G
- Emphasis on One #3: =A=B=C=D=E+F=G
- Emphasis on Two #3: =A=B=C=D=E+F+G
- Emphasis on Each: +A+B+C+D+E+F+G
- Unknown versus Each: -A-B-C-D-E-F-G
- Emphasis on Three #1: +A+B+C=D=E=F=G
- Emphasis on Three #2: =A=B=C=D+E+F+G
- Emphasis on Three #3: =A=B+C+D+E=F=G
- Extreme Relativism: ==A==B==C==D==E==F==G

II. Archetypal Motion Picture Premise Profiles:

Glowing for Two #1a: ++A++B-C-D-E-F-G
Glowing for Two #1b: +++A+B-C-D-E-F-G
Glowing for One vs. All Else #1a: +++A-D-E-G
Glowing for One vs. All Else #1b: ++A-C-D-F-G
Suppressing Three #1a: -A-B-C+F++G
Suppressing Three #1b: -A-B-C++D+E
Suppressing Three #2a: +A+B+C+D-E-F-G
Promoting Three #1a: +A++B+C-F
Suppressing Three #2b: +A++B+++C-E-F-G
Promoting Three #1b: ++A+B+C=D=E
Promoting Three #2: -C+E+F+G
Promoting Three #3: -A-B-C-D+E+F+G
Suppressing Three #3: +A+B+C+D-E-F-G
Glowing for Two #2: +++E+++F
Glowing for Two #3: -A-B=C=D+E+F
Glowing for One vs. All Else #2: -A+++B-C-D-E-F-G
Extreme Relativism: ==A==B==C==D==E==F==G
Basic Relativism: =A=B=C=D=E=F=G
Unknown versus Each: -A-B-C-D-E-F-G
Unknown versus Each and Every: -&-&-&-&-&-&-

III. Archetypal Interpersonal Energy Relation Profiles:

Love Between Two #1: A&G
Love Between Two #2: A&B
Love Between Two #3: C&D
Love Between Two #4: E&F
Love Between Two #5: B&G
Love Between Two #6: C&G
Hatred Between Two #1: A-vs.-G
Hatred Between Two #2: A-vs.-B
Hatred Between Two #3: C-vs.-D
Hatred Between Two #4: E-vs.-F
Hatred Between Two #5: B-vs.-G
Hatred Between Two #6: C-vs.-G
Apathy: -&-&-
Ambiguity: A&/orB&/orC&/orD&/orE&/orF&/orG
Transcendence: A/B/C/D/E/F/G/All/None/Hybrids

IV. Archetypal Rating Systems:

From least recommended to most recommended:

- 5 Integers: *, **, ***, ****, and *****

- 9 Half Steps: *, *1/2, **, **1/2, ***, ***1/2, ****, ****1/2, and *****

- 7 Half Steps: *, *1/2, **, **1/2, ***, ***1/2, and ****

- 10 Integers: 1, 2, 3, 4, 5, 6, 7, 8, 9, 10

V. Sample Rating Tendency Possibilities At Intersections

Mindset +A+B-C-D-E-F-G
meets
Premise ++A++B-C-D-E-F-G and
Energy Relation A&B:

5 Integers Rating: *****
9 Half Steps Rating: *****
7 Half Steps Rating: ****
10 Integers Rating: 10
...

Mindset +A+B-C-D-E-F-G
meets
Premise -A-B=C=D+E+F and
Energy Relation A&B:

5 Integers Rating: *
9 Half Steps Rating: *
7 Half Steps Rating: *
10 Integers Rating: 1
...

Mindset +A-B-C-D-E-F-G
meets
Premise -A+++B-C-D-E-F-G and
Energy Relation A-vs.-B:

5 Integers Rating: *
9 Half Steps Rating: *
7 Half Steps Rating: *
10 Integers Rating: 1
...

...

Mindset +A+B+C=D=E=F=G
meets
Premise +A+B+C+D-E-F-G and
Energy Relation C&D:

5 Integers Rating: ****
9 Half Steps Rating: ****
7 Half Steps Rating: ***1/2
10 Integers Rating: 8

...

Mindset -A+B-C-D-E-F-G
meets
Premise +A++B+++C-E-F-G and
Energy Relation A-vs.-B:

5 Integers Rating: *
9 Half Steps Rating: *
7 Half Steps Rating: *
10 Integers Rating: 1

...

Mindset -A+B+C-D-E-F-G
meets
Premise -A-B-C++D+E and
Energy Relation C-vs.-D:

5 Integers Rating: *
9 Half Steps Rating: *
7 Half Steps Rating: *
10 Integers Rating: 1

...

...

Mindset +A+B-C-D-E-F-G meets
Premise -&-&-&-&-&-&- and
Energy Relation B-vs.-G:

5 Integers Rating: ★★★★
9 Half Steps Rating: ★★★★1/2
7 Half Steps Rating: ★★★1/2
10 Integers Rating: 9

...

Mindset -A+B-C-D-E-F+G meets
Premise -&-&-&-&-&-&- and
Energy Relation B&G:

5 Integers Rating: ★★★★★
9 Half Steps Rating: ★★★★★
7 Half Steps Rating: ★★★★
10 Integers Rating: 10

...

Mindset =A=B=C=D+E+F=G meets
Premise -A-B=C=D+E+F and
Energy Relation E&F:

5 Integers Rating: ★★★★★
9 Half Steps Rating: ★★★★★
7 Half Steps Rating: ★★★★
10 Integers Rating: 10

...

...

Mindset =A=B+C=D=E=F+G meets
Premise +A++B+++C-E-F-G and
Energy Relation C-vs.-G:

5 Integers Rating: Trending with ** to ***** range.
9 Half Steps Rating: Trending with ** to ***** range.
7 Half Steps Rating: Tendency of *1/2 to **** range.
10 Integers Rating: Much spanning the 2 to 10 range.

...

Mindset =A=B+C=D=E=F+G meets
Premise +A++B+++C-E-F-G and
Energy Relation C&G:

5 Integers Rating: Polarizing heavily... * and *****
9 Half Steps Rating: Polarizing heavily... * and *****
7 Half Steps Rating: Polarizing heavily... * and ****
10 Integers Rating: Polarizing heavily... 1 and 10
...

Mindset =A=B=C=D+E+F=G meets
Premise -A-B=C=D+E+F and
Energy Relation E-vs.-F:

5 Integers Rating: **
9 Half Steps Rating: **
7 Half Steps Rating: **
10 Integers Rating: 4

...

...

Mindset -A-B-C-D=E=F+G meets
Premise +++A-D-E-G and
Energy Relation A-vs.-G:

5 Integers Rating: *
9 Half Steps Rating: *
7 Half Steps Rating: *
10 Integers Rating: 1

...

Mindset -A-B-C-D=E=F+G meets
Premise +++A-D-E-G and
Energy Relation A&G:

5 Integers Rating: *
9 Half Steps Rating: *
7 Half Steps Rating: *
10 Integers Rating: 1

...

Mindset +A+B+C+D+E+F+G meets
Premise -A-B-C-D-E-F-G and
Energy Relation A&/orB&/orC&/orD&/orE&/orF&/orG:

5 Integers Rating: *
9 Half Steps Rating: *
7 Half Steps Rating: *
10 Integers Rating: 1

...

...

Mindset -A-B-C-D-E-F-G meets
Premise -A-B-C-D-E-F-G and
Energy Relation A/B/C/D/E/F/G/All/None/Hybrids:

5 Integers Rating: *****
9 Half Steps Rating: *****
7 Half Steps Rating: ****
10 Integers Rating: 10

...

A Hybrid of Each of the Mindsets Described meets
A Hybrid of Each of the Premises Described and
A Hybrid of Each of the Energy Relations Described:

5 Integers Rating: Unobtainable/Unascertainable
9 Half Steps Rating: Unobtainable/Unascertainable
7 Half Steps Rating: Unobtainable/Unascertainable
10 Integers Rating: Unobtainable/Unascertainable
(with respect to the modeling choices in this chapter)
...

Mindset -A-B-C-D-E-F-G meets
A Hybrid of Each of the Premises Described and
A Hybrid of Each of the Energy Relations Described:

5 Integers Rating: Gravitating toward ****
9 Half Steps Rating: Gravitating toward ****^
7 Half Steps Rating: Gravitating toward ***
10 Integers Rating: Gravitating toward 8

...

...

Mindset =A=B=C=D=E=F=G meets
Premise -A-B-C-D+E+F+G and
Energy Relation -&-&-:

5 Integers Rating: **
9 Half Steps Rating: **
7 Half Steps Rating: *1/2
10 Integers Rating: oscillation between 3 and 4

...

A Hybrid of Each of the Described Mindsets meets
Premise +A++B+++C-E-F-G and
Energy Relation A&/orB&/orC&/orD&/orE&/orF&/orG:

5 Integers Rating: oscillation between * and ****
9 Half Steps Rating: oscillation between *1/2 & ****
7 Half Steps Rating: oscillation between *1/2 & ***
10 Integers Rating: oscillation between 2 and 9

...

A Hybrid of Each of the Described Mindsets meets
Premise ==A==B==C==D==E==F==G and
Energy Relation A/B/C/D/E/F/G/All/None/Hybrids:

5 Integers Rating: ***
9 Half Steps Rating: ***
7 Half Steps Rating: **1/2
10 Integers Rating: oscillation between 5 and 6

...

...

BONUS BETWEEN CHAPTERS EIGHTEEN & NINETEEN

On May 24th, 2023 at 10:37 AM (U.S.) CDT I sent an electronic message to Tienling Chen, who is the person described on page 117 as the cc-line intended recipient whom the Internet indicated would not be able to receive that message via the e-mail address that I attempted to use for him. Between the time of April 9th, 2023 and the later message, he had informed me that he had stopped using that old e-mail address, yet I decided then and still have decided to decline to forward him the correspondence that he had missed evidently due to circa-Q2-of-2022-to-circa-Q1-of-2023 closure of that old address. I do not know at the time of composition of this book whether or not I will ever forward that message to him in any form, and neither do I know whether or not he will ever find the copies of my 11:59 PM Houston time April 9th, 2023 e-mail message included elsewhere in this book.

Something I do know, however, is that I am including here the following copy of an excerpt from the aforementioned 5/24/2023 10:37 AM CDT message:

One year ago, on May 24th, 2022, after considering options of seeking a few very specific additional Vajrayana empowerments beyond those which I had already officially received via True Buddha School, I decided rather than to seek them through that organization (of which we are in the regular sense both members) or through another officially Institutionally approved and consecrated lineage of fully worldly presence at the time, to pray directly to the eleven Buddhas involved, requesting that they could at any given time, with or without my knowing it, grant or withhold from granting to me the empowerments requested in connection with them.

CHAPTER NINETEEN:
SIMPLY ONE OF THE MULTITUDES
OF WAYS TO CONSIDER
FIVE OR MORE DIMENSIONS
OF REALITIES

IR^2 IR IR^2

... ^{204}Tl IR ^{176}Tl ...

IR^3 ....   IR   .... IR^3

...   .

^{60}Co ^{90}Sr ^{90}Sr ^{60}Co

...   .

^{2}H ^{2}H ^{2}H ^{2}H ^{2}H ^{2}H ^{2}H

^{2}H ^{3}H ^{2}H ^{3}H ^{2}H ^{3}H ^{2}H

10   *****   ****   *****

IR^5

...   .

^{235}U ^{235}U ^{235}U ^{235}U ^{235}U ^{235}U ^{235}U

....

IR^4

....

^{239}Pu ^{239}Pu ^{239}Pu ^{239}Pu

... ^{239}Pu ^{239}Pu ^{239}Pu ...

.... IR^4 IR^5 IR^4 ....

9   10   9   10   10   *   1   ***1/2   10

...^{3}H IR^7 ^{3}H... IR^8 ...^{3}H IR^7 ^{3}H... ... ...

^{90}Sr ^{3}H ^{3}H ^{90}Sr ^{3}H ^{60}Co ^{3}H ^{143}Te ^{3}H

^{3}H ^{3}H ^{3}H ^{3}H ^{3}H ^{3}H ^{90}Sr ^{90}Sr ^{90}Sr ^{60}Co ^{3}H ^{143}Te

^{142}Te ^{90}Sr ^{90}Sr ^{90}Sr ^{90}Sr ^{90}Sr ^{90}Sr

.... IR^9 ....

APPENDIX I

<u>Part I: An April 18th, 2023 Outtake Synopsis for *All Things under and over the Sun and Stars: Enigmas in Various Stages*</u>

Settlers on another planet rebel against the rest of the human race and eventually seek to subjugate the entirety of humanity, resulting in fourth-millennium warfare between The Planet of New Gwalintu and The Planet Earth. After extreme developments, escalations, and what seems to be a decisive resolution, that resolution proves to be an illusion, and scenes shift to an alternate version of the early 21st century. There The United States of America and The Russian Federation exist only in fiction, whereas The United States of Philsatlan and The Democratic Republic of Belarus are superpowers that exist in historical reality and geopolitical current events. Peacetime entry into the age of nuclear weapons in their world leads to grave concern, a grand conference, and an assassination plot. After climactic drama and its aftermath unfold on Earth in that universe, things jump to yet another alternate universe, where 51st-century space exploration leads to revelations tying together much of what happened before and what might happen next, answering many questions and posing a multitude of new ones.

<u>Part II:</u> Please note that the October 5th reference on page 114 of *The Dimetrodons, the Dorians, and the Modern World: Revised Edition* and the October 6th reference on page 115 of that book both involve an occurrence that happend on 10/5 as measured from Texas and 10/6 as measured from Sri Lanka.

Part III: A Bibliography of Recommended Further Readings

[Perhaps Unknown Author(s), sometimes presumed to have been Shakyamuni Buddha and/or Member(s) of the Early Buddhist Sangha.] *The Dhammapada*. (any of various manifestations, some of which are in the public domain, others of which are not yet in the public domain). {A notable example is the English translation by The Venerable Balangoda Ananda Maitreya with revision by Rose Kramer, which has at least two published manifestations.)

Ouspensky, P.D. *Tertium Organum*. 2nd ed. in English. Translated by Nicholas Bessaraboff and Claude Bragdon. (any of various manifestations; in the public domain). 1920.

Ouspensky, P.D. *Tertium Organum*. 3rd American Edition, Authorized and Revised. Translated by Nicholas Bessaraboff and Claude Bragdon. New York: Alfred A. Knopf. 1945.

Thrangu Rinpoche, Khenchen. *The Five Buddha Families and the Eight Consciousnesses*. Translated by Peter Alan Roberts. Glastonbury: Namo Buddha Publications. 2013.

Various Writers, Some of Them Somewhat Known (e.g., Julia Ward Howe) and Others of Them Rather Unknown. Lyrics for variations of songs of the titles, "John Brown's Song," "John Brown's Body," and/or "The Battle Hymn of the Republic." Portions of the period from circa 1861 onward.

APPENDIX II

Consider the following chronology, drawn from multiple sources, including, at times, comparison and contrast between Wikipedia (i.e., wikipedia.org), Conservapedia (i.e., conservapedia.org), and general communications:

Part I: Antiquity to 1611:
Differing expressions emerge about multiple scientific and religious issues, not the least of which were/are:

- Identities, Interpretations, and Demarcations of what people would point toward with the phrase "The Ten Commandments" (e.g., whether to use *Exodus* 20:2-17 or to use *Exodus* 20:3-17)
- Identities and Demarcations of what people point toward with chapters of *Psalms* (e.g., how some use the nineteenth chapter to indicate what others would indicate as being the eighteenth chapter) (Cross reference: similarities and differences between *The Duoay-Rheims Bible* and *The King James Version Bible*)
- Whether there is only one correct namable sect of only one correct namable religion, whether there are two or more correct namable sects, whether there are two or more correct namable religions, whether a person has transformability between religions, whether unnamable sects and/or unnamable religions would be superior to namable ones, whether at least one namable sect and/or at least one namable religion would be superior to unnamable ones, whether there exists room enough for different religions and sects to coexist or not, and the degrees to which killing in the name of politics, political science, religion, science, and/or economics could, would, and/or should ever be justified, justifiable, necessary and/or sufficient

Part II: 1561-1763

<u>1561:</u> Lightning resulted in a fire that destroyed the steeple of [Old] St. Paul's Cathedral of London, England.

<u>1642:</u> War pitting English against English began (relative to developments considered key by many).

<u>1646:</u> Some considered civil war between English to have temporarily ended, whereas others considered it to have continued.

<u>1648:</u>
- Outbreak of the Second English Civil War (as considered by some, of 1648-1651) and significant changes to The English Civil War (as considered by others, of 1642-1651) occurred.
- Oliver Cromwell and associated persons instigated for England to place King Charles I on trial.

<u>1649:</u>
- January 30: Execution of King Charles I by orders of the Cromwell-and-associates-led Government of England occurred.
- September: Siege of Drogheda in Ireland happened.

<u>1650:</u> Escalation of Conflict Between King Charles II (who had traveled to Scotland and found support there) and Oliver Cromwell unfolded.

<u>1651:</u> Some consider that it marked the end of The English Civil War, whereas others considered that The First English Civil War was during 1642-1646 and The Second English Civil War happened in the 1648-1651 range; War pitting English against English ended (relative to developments considered key by many historians).

<u>1756:</u> In Calcutta, India, on June 20, the Nawab of Bengal arranged to place English prisoners into an extreme situation of detention and confinement; few survived.

<u>1756-1763:</u> War occured in many places in Europe and North America, largely involving boundary disputes.

....

<u>Part III: October 10, 1911 to July 4, 2023</u>

<u>Portions of 1911-1930 from select vantage points:</u>

- The Xinhua Revolution started on October 10, 1911.
- The Philadelphia Athletics won the Major League Baseball World Series from October 14-26, 1911.
- The first known instances of the human use of combat airplane missions took place from October 23 to November 1, 1911 between the Kingdom of Italy and the Ottoman Empire (in what some refer to as the Turco-Italian War), changing the future trajectories for what would become Turkey, Italy, and many remainders of the realities.
- The P.D. Ouspensky book first written and published in the Russian language and later translated into other languages to be known as *Tertium Organum* reached a variety of audiences in 1912.
- The Great War, which would later become known as World War I, happened 1914-1918.
- After the Bolshevik revolution of 1917, Soviets declared a ban of multiple characters of the previous version of the Russian alphabet, effective 1918.
- The Russian Civil War spanned portions of 1917-1923 and featured a 1922 merger of multiple Soviet states into the U.S.S.R. version of Russia.

Portions of 1913-1947 from select vantage points:

- On July 14, 1913, Leslie Lynch King, Jr. was born. In 1917 he became known most often as Gerald Rudolff Ford, Jr., though formalization of the name change waited until 1935.
- English translations of *A New Model of the Universe* by P.D. Ouspensky became published: first edition on April 24th, 1931 & second edition, revised, in September 1934 {as reported by a January 1946 reprint of the reset-via-new-plates August 1943 print of the Second Edition, Revised, which had Alfred A. Knopf as its publisher}.
- World War II happened 1939-1945.
- Finland had a separate 1939-1940 war with the U.S.S.R.
- Also, Finland achieved a draw in a sometimes mixed and eventually fully separate war vs. the U.S.S.R. 1941-1944 or 1941-1947 (according to some of the differing interpretations). Although Finland formed a de facto temporary alliance with the Axis Powers against the Allied Powers in portions of 1941-1944, Allied leaders in 1942 declared that Finland was in no way hostile to the Allies, rather engaged in a war with the Union of Soviet Socialist Republics *as an aside* to the main European Theatre of Action in World War II. Despite making a separate armistice agreement with the Soviet union in September 1944, the Fins were still at risk of extra territorial concessions to the Soviets in connection with the then-ongoing conflicts between the U.S.S.R. and other nations. After VE Day in May 1945 and V-J Day in August 1945, the alignments of Soviet activities within the geopolitical landscape shifted. *Finland declined Marshall Plan aid in 1947 to help preserve Finnish autonomy.*
- P.D. Ouspensky (born 1878) died in 1947.

<u>1951:</u> *The Day the Earth Stood Still* reached movie audiences and achieved success.

<u>1952:</u>
- November 1: The United States of America successfully detonated a hydrogen bomb.
- November 4: There arrived the culmination of how Dwight D. Eisenhower and Adlai Stevenson II competed in a landslide U.S. presidential election.
- November 16: The United States of America reported the aforementioned thermonuclear detonation test from fifteen days earlier to have been the first known instance of a human-created hydrogen bomb explosion.

<u>1953:</u>
- *Childhood's End* (novel) by Arthur C. Clarke reached its first publication and achieved both commercial and critical success.
- "The Nine Billion Names of God" (short story) by Arthur C. Clarke reached publication and success.

<u>1960:</u>
- *Carthage in Flames* reached movie audiences.
- John F. Kennedy and Richard Nixon competed in a controversial U.S. presidential election.

<u>1961:</u> The New York Yankees won the Major League Baseball World Series.

<u>1962:</u> *The Manchurian Candidate* starring Frank Sinatra, Laurence Harvey, and Janet Leigh reached movie theaters, sending very mixed possibilities of messages and interpretations amid achieving critical success.

<u>1963:</u>
- In October, broadcasting as part of *The Outer Limits* series was the TV episode "The Man Who Was Never Born."
- In November, U.S. President John F. Kennedy (b. 1917) and U.S. citizen Lee Harvey Oswald (b. 1939) both died. (Reports by many indicated that the latter acted as the lone assassin of the former, though that characterization has been disputed by some {on account of alleged peculiarites and/or perceived impossibilities in the physics seemingly implied by some intuitive viewings of the Zapruder film footage in comparison and contrast with the officially reported explanations, as well as by other factors}. Lee Harvey Oswald received mortal wounding from a gunshot wound delivered by Jack Ruby right in front of law enforcement while in police custody.)

<u>1967:</u>
- Jack Ruby (b. 1911) (assassin of Lee Harvey Oswald) died.
- "White Rabbit" (audio recording song) by Jefferson Airplane became released to the public.
- "The City on the Edge of Forever" (*Star Trek* episode written by Harlan Ellison and Gene Roddenberry) first became broadcast on television.

<u>1969:</u> The pilot episode of *Night Gallery* had its initial broadcast on television in November.

<u>1971:</u>
- *Rules for Radicals* by Saul Alinsky reached its first publication. It would go on to prove extremely controversial.
- Edith Tolkien (b. 1889) died in November.

1972:
- The Dallas Cowboys won Super Bowl VI early in the year.
- David Carradine starred in early episodes of the *Kung Fu* television series. Broadcasts reached television audiences and proved commercially and critically successful.
- Saul Alinsky (b. 1909) died on June 12.
- The controversial Watergate burglary happened on June 17, eventually leading to the Watergate political scandal.
- Initial filming began for a project that Bruce Lee, Dan Inosanto, Kareem Abdul Jabbar, and others would collaborate on and which Lee envisioned to use martial arts as a metaphor for the challenges related to Death Itself. That project would not reach movie theaters until several years later.
- Richard Nixon and George McGovern competed in a landslide U.S. presidential election.

1973:
- The Miami Dolphins won Super Bowl VII.
- Secretariat won the Triple Crown in thoroughbred racing.
- Jeet Kune Do Founder Bruce Lee (b. 1940) reportedly died, though some have generated extravagant theories that he might have gone into hiding while strategically faking his death.
- *Enter the Dragon* starred Bruce Lee and reached movie theaters, achieving for Lee much (presumably posthumous) crossover Hong Kong-to-Hollywood success.
- J.R.R. Tolkien (b. 1892) died in September.
- The Oakland Athletics won the Major League Baseball World Series in October.

<u>1974:</u>
- The Miami Dolphins won Super Bowl VIII early in the year.
- President Richard Nixon resigned in August.
- Gerald Rudolff Ford, Jr. became President of the United States upon Nixon's resignation, and he became generally referred to as President Gerald Ford.

<u>1975:</u>
- Large numbers of citizens of The United States and multiple other countries frantically withdrew as many of themselves as they could on an emergency basis from Vietnam.
- *The Basement Tapes* (audio recording album) by Bob Dylan and The Band became released to the public.
- *Physical Graffiti* (audio recording album) by Led Zeppelin became released to the public.

<u>1979:</u> Iran underwent a process of regime change.

<u>1981:</u>
- The Oakland Raiders won Super Bowl XV early in the year.
- *Raiders of the Lost Ark* reached movie theaters and achieved both critical and commercial success in middle portions of the year.

<u>1982:</u>
- The St. Louis Cardinals won the Major League Baseball World Series.
- *2010: Odyssey Two* (novel) by Arthur C. Clarke became published and distributed for the first time.

<u>1984:</u> The Los Angeles Raiders won Super Bowl XVIII.

<u>1989:</u>
- *The Seven Habits of Highly Effective People* by Stephen R. Covey reached its first publication.
- The Oakland Athletics won the Major League Baseball World Series.

<u>1990:</u> The Cincinnati Reds won the Major League Baseball World Series.

<u>1991:</u> On March 25, the 63rd Academy Awards ceremony took place, honoring motion picture films.

<u>1989-1993:</u> The Vatican's 1989 conflict with Madonna Louise Ciccone over the music video "Like a Prayer" (which some may interpret as a boiling over of years of tension between The Vatican and Ms. Ciccone over how Ms. Ciccone would professionally use the mononym Madonna as a popular singer, whereas the mononym Madonna is one of the main ways to refer to The Virgin Mary), became debatable as to how and to what degree, if any, it might relate to the 1992 publication of the Duran Duran song "Ordinary World" and the 1993 publication of the Arthur C. Clarke novel *The Hammer of God*.

<u>1994:</u>
- President Richard Nixon (b. 1913) died.
- *When Fallen Angels Fly* (audio recording album) by Patty Loveless reached its first publication in August.
- Major League Baseball experienced negotiation breakdowns resulting in a strike which began in August and wiped out the remainder of that regular season, causing what might otherwise have been the 1994 MLB World Series to not occur.

<u>1995:</u>
- The 1994-1995 Major League Baseball Strike ended in April.
- The Houston Rockets won the NBA Finals in June.

<u>2000-2001:</u>
- George W. Bush and Al Gore competed in the highly controversial 2000 U.S. presidential election.
- September 11, 2001 became one of the most controversial and shocking dates in history, including the Al Quaeda hijacking of four large jet airplanes. Ensuing were the crashing of three out of the four into Osama bin Laden-designated strategic targets in the United States, whereas one out of the four became diverted by passengers who overcame their plane's hijackers. In the aftermath, air travel in the U.S. became grounded for several days, people scoured the Internet for real and alleged prophecies (including how Nostradamus became especially popular for a while as a search engine input), and large numbers of Americans started to binge-watch news programming on television much more than they had previously done.
- In the Fourth Quarter, Afghanistan underwent regime change, largely as U.S. retaliation for the 9/11/2001 attacks, in connection with the Taliban having enabled many of the Al Quaeda activites that led to those attacks.

<u>2002:</u>
- "Courtesy of the Red, White and Blue (The Angry American)" (audio recording song) by Toby Keith became released to the public.
- *American IV: The Man Comes Around* (audio recording album) by Johnny Cash became released to the public.

2003:

- Iraq underwent a process of regime change.
- June Carter Cash (b. 1929) died.
- Johnny Cash (b. 1932) died.

2004:

- Ferrari thoroughly dominated the Formula One season, winning the constructor's championship by a huge point margin.
- George W. Bush and John Kerry had a moderately-competitive U.S. presidential election.
- The Boston Red Sox won the Major League Baseball World Series.
- December 26: A tsunami killed a number of people estimated to have been greater than 200,000 in total, emanating from a Richter-9.1-9.3-magnitude earthquake off the coast of Indonesia.

2005:

- In the late-June-to-early-July period a drink featuring the four-letter name "HALA" on its containers appeared for sale in at least one Asian-themed grocery store in the International District in Houston, Texas.
- The 31st G8 Summit occurred July 6-8.
- *War of the Worlds* (directed by Steven Spielberg and starring Tom Cruise) reached movie theaters in July.

2006:

- December 26: President Gerald Ford died.
- December 31: People acting by authority of the government of Iraq executed that country's former ruler Saddam Hussein (b. 1937).

2008-2009: A catastrophic global financial crisis occurred.

2011:
- Osama bin Laden (b. 1957) died as part of an American action following up on the aftermath of the events of September 11, 2001 and related impetus.
- CERN's Antihydrogen Laser Physics Apparatus (ALPHA) project sustained antihydrogen atoms for over sixteen minutes.
- Japan won the FIFA Women's World Cup Soccer competition.
- The St. Louis Cardinals won the Major League Baseball World Series.

2012:
- *Ameritopia: The Unmaking of America* by Mark R. Levin reaches publication. It would prove controversial within some spheres of influence.
- Stephen R. Covey died.
- The opening ceremony of The Summer Olympics included The United Kingdom featuring the main studio recording of the famous-since-1973 song "Eclipse" by Pink Floyd.

2014: Germany won the FIFA World Cup Men's Soccer competition.

2015: A Richter-8.3-to-8.4-magnitude earthquake struck offshore near South America on September 16.

2016:
- The Denver Broncos won Super Bowl 50.
- The Cleveland Cavaliers won the NBA Finals.
- The Chicago Cubs won the Major League Baseball World Series.
- Donald Trump and Hillary Clinton competed in a controversial U.S. presidential election.

2017:

- January 1: In Istanbul, Turkey, a nightclub shooting happened.
- January 19: A constitutional crisis escalated in Gambia as part of the wake of a contested 2016 presidential election.
- February 5: The New England Patriots won Super Bowl LI by a score of 34-28 in overtime over The Atlanta Falcons.
- March 22: The United States won The World Baseball Classic.
- August 17: Scientists reported observation of two neutron stars colliding.

2018:

- February 4: The Philadelphia Eagles won Super Bowl LII by a score of 41-33 over The New England Patriots.
- December 13: In Ankara, Turkey, two trains collided.

2019:

- *Captain Marvel* (starring Brie Larson) first reached film enthusiasts via movie theaters and other means of delivery.
- The First Impeachment of President Trump by the United States House of Representatives happened.
- *The Irishman* (directed by Martin Scorsese) first reached moviegoers through a variety of means.

2020:

- The First Impeachment Trial of President Trump resulted in acquittal by the United States Senate.
- *Wonder Woman 1984* reached movie viewers via multiple means.
- Joe Biden and Donald Trump competed in a controversial U.S. presidential election.

2021:

- On January 6, centuries-long escalations of North American political tensions reached a flashpoint, exploding into a highly destructive U.S. Capitol riot.
- The Second Impeachment Trial of President Donald Trump resulted in acquittal in February.
- The United States Armed Forces completed the remainder of withdrawal from Afghanistan in August, in connection with the 2001-2021 hostilities, including the Q4 2001 overthrow of the Taliban from control, the subsequent insurgency, and the combined aftermath of millennia of conflicts in that region of Earth.
- Entertainment industry professional Betty White (b. 1922) died on December 31.

2022:

- National Football League (NFL) professional Dan Reeves (b. 1944) died on January 1.
- *Don't Worry Darling* (directed by Olivia Wilde) and *Black Adam* (starring Dwayne Johnson) reached movie theaters, Argentina won the FIFA World Cup Men's Soccer competition, and The Houston Astros won the Major League Baseball World Series.
- *Avatar: The Way of Water* (directed by James Cameron) reached movie theaters in December.

2023:

- February 7: Missouri executed Leonard Taylor (b. 1964).
- February 12: The Kansas City Chiefs won Super Bowl LVII (score: 38-35 vs. The Philadelphia Eagles).
- May 2: Movie studios and the Writers Guild of America (WGA) had a communication breakdown, resulting in the start of a writers' strike.
- May 29: Nepalis dominated the 42k & 70k competition sections of the 19th Everest Marathon.

APPENDIX III: A more complete portrait of April 9th, 2023 (as measured by U.S. Central Daylight Time) (omitting copies of the prior rounds of e-mails and the e-mail attachments referenced by any older and newer rounds in the thread of back-and-forth messages):
------forwarded message: 8:31 PM from Blair to Berry------
Michael Berry, Emily Bull, and/or other show personnel,

Thank you for sharing the photograph and arranging the related events and broadcasts.

Although I sent the check via USPS Priority Mail and their system indicated that it arrived (to the show address you indicated via e-mail) well ahead of the initial transfer of checks to Joyce (reference: see the attached pdf), the check #2006 that I sent (in the amount of $160) has not shown as processed by the bank as of a few minutes ago (in my online account access).

Please verify that you received that package successfully and took the appropriate steps.

Of course, if Joyce received the check and has chosen to either delay or avoid depositing it while keeping my bank account information reasonably confidential, then that legally and ethically could be her prerogative as the intended recipient; however, if anyone in the actual chain of custody obstructed her receipt of the check, then that would constitute mishandling.

Do you have any comments on this?

Regards,

Maurice J. Blair

-------{first portion (of two) other than the copies of prior rounds of the thread of communications}----------------------

Are you serious? Your check didn't get deposited by Joyce? "Mishandling"? Cancel the check with the bank. We forwarded all checks & can't help what Joyce does after that. We received lots of checks & sent them all on. If your check didnt cash, nobody stole your money. Cancel your check with the bank. You are blocked. You crank.

---------{second portion (of two) other than the copies of prior rounds of the thread of communications}--------------

--

Emily Bull
Assistant to

The Czar of Talk

Love listening to the Czar? Get the daily recap of the show, plus more!
Go to www.MichaelBerryShow.com & click on "Join Michael's Email List". You'll get a daily afternoon email with some of our best stuff. We'll never sell or share your email. Pinky swear.

Revisiting The Response That Already Appeared in This Book in Portions of Intermission One

- This time redacting the two e-mail addresses in a slightly different manner, and with a now more complete context.
- It is unknown to me at the time of composition of this book whether Michael Berry of *The Michael Berry Show* is aware of how deeply his assistant Emily Bull had offended me.
- Also, although I let go of taking offense and showed a mixture of forgiveness and rebuke, as time went by and there came no direct response, my memories and love of seeking good outcomes to situations led to tuning in to how the Reverend Al Sharpton said on at least one TV news broadcast appearance many years ago that forgiveness can and should often go together with dishing out justified punishment to those who perpetrate harm.
- Please bear in mind that I have heard at least one other minister type of person present via a broadcast the idea that even when people receive forgiveness, consequences can and often do happen, and that there are plenty of scriptural bases for that view to be applicable, at least some significant portion of the time.
- As with many interactions, I am less concerned with the basic worldly level of actions and consequences than with the Absolute and Awesome Levels of Ultimate Reality (sometimes referred to as "God," et cetera) that have demonstrated to me time and again that He/It/They/&cetera can and do perform the unthinkable unto the human experience of reality, as I have experienced the hitherto-unthinkable repeatedly on occasion in my life (whether viewed as paranormal or something else).

sent at 11:59 PM U.S. CDT on April 9th, 2023 from Blair to Berry's e-mail address (with the two Blair e-mail addresses that had appeared in the original text redacted from this copy)

Michael, Emily, et al at The Michael Berry Show,

To clarify, I was not aiming at accusing your show specifically of theft (though there was a slight suspicion of a strategic temporary delay by your show and a larger suspicion of it not ever arriving at your show's address). *I was* aiming at stating that *anyone in the chain* (including neighbors who might have intercepted it - or even - as a stretch - at least one rogue member of USPS personnel - if cognizant of portions of my 2019, 2021, and 2022 appearances on your show and influenced by any motive(s) to attempt to prevent it from arriving) if mishandling it could be causing a problem!

I do not need to cancel the check; $160 is a small amount to keep as part of the padding with that account. So you dared to try to tell me to cancel the check; I am declining to cancel that check!!

YOU OVERREACTED!!!!!

You indicated blocking my [first redaction from what would go on to be forwarded via S.R.P.C. & M.J.B.] address and I do not know whether you had blocked this [second redaction from what would go on to be forwarded via S.R.P.C. & M.J.B.] address prior to my attempting to send this message to you, yet there is a chance that you will become aware of this message even if it does not electronically arrive at your inbox,

{256}

especially due to the use of the cc line and other factors.

Yes, you can choose to continue to take offense from my recent inquiry and choose to keep me blocked and think of me as a crank for daring to even bring this up with you; you have that liberty in this country, but after the several hours of caller time that you granted me (for which I am still thankful and will likely continue to be thankful for in the future, even if your show chooses to make that block permanent and refrains from future reversal of it) spread out over time I had not expected you to read my message with such a spirit of irritability and hostility. Feel free to someday unblock me or to refrain from doing so; either way I am choosing today, to forgive you, Emily, Michael, and/or whoever else is responsible for this recent over-the-top harshness that I believe constitutes a gross overreaction on your part.

Also, I hereby pray for your ability to discern better when someone sends you a message in a benevolent spirit without realizing that its compositional structure might otherwise trigger you, and I pray for my ability to better anticipate how to refrain from sending a message that might trigger someone with the sensitivities that you exhibited - or to compose such a message in a way less likely to trigger someone with those sensitivities. In a future similar situation with someone else I might consider writing something more like, "That check I sent to you for you to forward to its recipient did not get deposited yet as far as I know, and it's been quite a while. Would you please confirm that you actually received it? I hope all is well with this situation."

To shine lights on any and all and every degree to which this communication breakdown reflects negatively, positively, or both on you and/or me and/or anyone else, I am choosing to include some recipients we to a degree mutually respect on the cc line (though the first three recipients are probably much more familiar with you than they are with me), plus a person I know rather well and of whom you and the first three cc line recipients were probably not previously aware.

Best Wishes to Have a Blessed Remainder of April 2023, including to whatever degrees the giving and the receiving of the rebuke of the wise may be part of this!

Maurice J. Blair

P.S., Whichever way this proceeds, I look to the past, present, and future with no regret toward our interactions and I truly believe it will become part of the good in the long run. I have let go of taking offense from this set of communications, and you can consider whether in a given minute or decade to also let go of taking offense from this set of communications.

Note to the readers of this book: As a reminder, the check #2006 did indeed clear the bank on the next day, which was April 10th, 2023, per official financial account records.

APPENDIX IV

A More Complete View of May 24th, 2023 communications

---------- a complete transcript of the Facebook Messenger message sent from Maurice James Blair to Tienling Chen at 10:37 AM CDT on 5/24/2023 -------------

Tienling, Happy Hermes Trismegistus Day! Although Hermes Trismegistus is somewhat beyond having any specific official relationship with True Buddha School, some would consider him and GM Lu Sheng-Yen to both be very much involved with soteriology. On another note, although I am aware that you and John Lin had some type of falling out with each other many years ago and am uncertain whether he and you ever made amends with each other since, I will let you know that I respect both him and you. That being said, I have tended to get along better with you - or at least perceived myself to get along better with you - as considered from a holistic perspective in the time from first meeting TBS personnel in 2002 up to now. Here is something somewhat peculiar: In a Dharma Class session type of setting, with John and his wife Vivian Lin as the main instructors, approximately in the range from the fourth quarter of 2003 to sometime 2004, they presented the idea that long ago there was a village where the people were so corrupt and terrible that Padmasambhava, in righteous indignation and out of a duty to unleash divine wrath, killed each and every resident of the village. Many years later, I found that at least some translators have shown in English the story to indicate that, in contrast with the aforementioned, Guru Rinpoche had killed all the males and achieved some type of transcendental merger with all of the remaining females in the village. Who knows how and why the oral tradition stories morph the way they do over the eons and why the written stories and translations wind up taking on the fixed - or somewhat fixed - forms that they do? On another note, at the risk of being attacked by you or by a high percentage of beings heavily involved with True Buddha School, though I believe it unlikely that such an attack will actually come to pass (and will likely have a reasonably enlightening resolution if and when it were to come to pass), I shall now reveal to you (and whoever might somehow look at this message, now or ever in the future, via surveillance or whatever else) something else: One year ago, on May 24th, 2022, after considering options of seeking a few very specific additional Vajrayana empowerments beyond those which I had already officially received via True Buddha School, I decided rather than to seek them through that organization (of which we are in the regular sense both members) or through another officially institutionally approved and consecrated lineage of fully worldly presence at the time, to pray directly to the eleven Buddhas involved, requesting that they could at any given time, with or without my knowing it, grant or withhold from granting to me the empowerments requested in connection with them. Although I decline to type here anywhere near a rather full narrative description of all that took place in the transitions from waking states to sleeping states to waking states after that prayer that evening and proceeding into early portions of the next day, suffice it to say that there were very strong dynamics open to interpretation of confirmation. I do on rare occasions chant the chosen formats of mantras in connection with those, with full faith in acting on whichever degree, if any, of empowerment through direct channels carrying forward the legacy of that experience. I am aware that GM Lu indicated that he has done similar things before as part of founding TBS and expanding its prowess, and, yes, I am aware that various TBS practitioners have warned that for people to attempt higher level practices outside of institutionally-approved-and-arranged lineage methods of empowerment could be extra dangerous; however, I also remember seeing somewhere an English translation of a passage in which Lu Sheng-Yen expressed that people using Zen mind could be capable of virtually anything, even, in cases extreme enough, actions like destroying a statue of him. On another note, shifting over to looking at the waters in the photograph you shared in the message you sent about 10 1/2 days ago, it reminds me that a few years ago or thereabouts I looked around and saw that there were no books listed available for sale on a major online retailer in connection with the title *The Waters of Oblivion*, though it seems one of the most obvious of possible book titles imaginable. Who knows what books - whether currently available for sale to the general public or not - might have that title... or which someday might emerge with that title? Regards, M.J.B.

------ Tienling Chen's response to Maurice James Blair on 5/24/2023 at 11:11 AM U.S. CDT ----------------------------

Well , I had a nice dinner with John Lin and some members in the temple a few weeks ago . And it was a good experience ! Thanks for caring ... ☺

APPENDIX V

A LIST OF THE UNIVERSITY OF TEXAS AT AUSTIN PERFORMING ARTS CENTER EVENTS FOR WHICH MAURICE JAMES BLAIR SERVED AS A STUDENT VOLUNTEER USHER:

10/26/2001: *The Magic Flute* (University of Texas at Austin Music School production of that Mozart opera)

11/10/2001: *Faust* (Austin Lyric Opera 1 production with Orchestra)

12/20/2001: *The Nutcracker* (Ballet Austin presenting traditional Christmas ballet and youth show of that Tchaikovsky suite)

01/18/2002: *A Streetcar Named Desire* (Austin Lyric Opera production with orchestra; music by Andre Previn, libretto by Philip Littell; based on play by Tennessee Williams)

02/02/2002: *South Pacific* (Touring Broadway musical of the beloved Rogers and Hammerstein classic)

03/15/2002: *The Rite of Spring* (Ballet Austin 3 production featuring Mills' choreography of Stravinsky's music)

03/17/2002: *The Rite of Spring* (Ballet Austin 3 production featuring Mills' choreography of Stravinsky's music)

APPENDIX VI: ADDITIONAL SUPPLEMENTAL PERSPECTIVES

The first two items below are footnotes to the preceding proceedings of the pages of this work as an entirety, whereas the third item onward are footnotes to the second item.

1. Several times over the span of many years, Maurice A.T. Blair told me about how, sometimes in the projects he was involved with professionally (in the U.S. military and/or with Raytheon), he and others had learned by experience that physical manifestations that would normally in tests exhibit equations like I^2R would at close distances and with high power sources suddenly exhibit those equations escalating into I^3R, I^4R, I^5R, etc.

2. A man named Victor Rivera mentioned to me in Austin, Texas in mid-2001 the idea that perhaps I should check out the book *Simple Abundance* sometime. Soon after that, I indeed checked out that Sarah Ban Breathnach book from a local library and read many parts of it - though not anywhere near the entirety of it - before checking it back in to that library. Early in the time of studying it and considering its perspectives in comparison and contrast with parental guidance, peer guidance, and other things, I quickly noticed a passage describing the idea that: 1) God will sometimes inspire a person with the potential to start a creative project, 2) The inspired person will need to decide whether to proceed with working on the project in earnest or to decline to work on the project in earnest, 3) God will then co-create with the person if the person chooses to proceed, whereas God will move along to inspire someone else with the assignment if

the person chooses to decline to proceed. Although I have not looked at that passage since mid-2001, and I typed the above synopsis of a minute fraction of one section of that book from a combination of memory and impressions (and some of my own way of conceptualizing and conveying multiple essences), I remember clearly and distinctly something highly noteworthy in relationship with that, which I shall now share with you. Although I had accumulated a large quantity of knowledge, experience, and insight from April 12, 2000 to the instant that I encountered the aforementioned passage in *Simple Abundance*, and although I had lightly considered the idea of creating a novel on multiple occasions during that span, it was not until I read and considered that passage that I started to seriously consider the possibility of creating a novel.

Here is where things get weird with that part of my personal history, weirder than they had previously been, that is. Instead of a normal author's mentality toward creating a novel, if such a thing or range can truthfully even be said to actually exist, I had a mentality that involved a critical mass of weirdness. Seems appropriate that it happened in Austin, doesn't it?! Well, back when I experienced great insights and some very mysterious expressions of sensory perception of spatiotemporal relations of audiovisuals and beyond in the second quarter of 2000, I soon started to view myself as having achieved spiritual and intellectual riches beyond my previous wildest dreams! I remembered the passage in *The Gospel According to St. Matthew*, as conveyed by many people over the years, of how "it can be harder for a rich person to enter the Kingdom of Heaven than for a camel to pass through the eye of a needle." (Cross reference: *Matthew* 19:23-24.) Also, I remembered the related idea that part of the way to achieve the seemingly impossible would be to - either literally or metaphorically - "give away everything one owns and depart from the old way of life to follow Jesus in a new way of life." (Cross reference: *Matthew* 19:21.) Although at the time I was not completely sold on the idea presented by psychiatrist Dr. Jenna Saul and by various Christian clergy and laeity over many years that God would exclusively interact with humans in the long run in a "choose His way by choosing Christianity or else choose The Highway to Hell" / "His Way or The Highway" manner, I had arrived at a belief that there had to be a major amount of legitimacy in Christianity, though uncertain exactly how high a percentage of reality might be best to render unto the Christian ideas of reality.

That combined with an enthusiasm for exploring further and an extreme enthusiasm for Clarke's *Space Odyssey* series {which had been part of the inspiration for my transformation from a) generalized spirituality (prior to mid-1998) to b) annihilationist atheism with a slight edge of agnosticism (much of mid-1998 to sometime December 1999) to c) thorough agnosticism (sometime December 1999 to sometime April 12, 2000 while in the middle of taking a Management Accounting examination in a Houston Community College classroom under the instruction of Mel McQueary) to d) wide-open esotericism (from the aforementioned instantaneous flash of going beyond on April 12, 2000), as different experiences and thoughts proceeded} {having read *2001: A Space Odyssey* circa mid-August 1997, *2010: Odyssey Two* circa late-August-to-mid-September 1997, *2061: Odyssey Three* circa mid-to-late-December 1997, and *3001: The Final Odyssey* circa May 1998}.

That openness toward Christianity - tempered by a healthy skepticism toward jumping into extreme sectarian dogmatism within it - combined with an enthusiasm for exploring further while treasuring insights from Clarke's *Space Odyssey* series, Clarke and Kubrick's related co-created *2001: A Space Odyssey* novel-and-book-each-based-partway-on-the-other-and-each-partway-the-basis-of-the-other projects, plus a degree of love for the novel *2010: Odyssey Two* and its subsequent big-screen adaptation *2010: The Year We Make Contact* all combined into what many might have considered an incredibly hare-brained performance art concept: I set out after a while onto a plan to a) possibly create a magnum opus of a novel on a semianonymous basis, b) write a disappearance will and testament leaving my previous possessions to various people, c) transfer copies of that novel to various institutions while granting them the rights to reprint it unaltered, and d) join the French Foreign Legion to adopt a new, anonymous life and become either very difficult or next-to-impossible for anyone who previously knew me to somehow find. Also, I planned to pull off the stunt in the year 2010, to parallel patterns in the novels and movies, as an act of extreme performance art, and as a way to carry out a very literal intepretation of fulfilling what Jesus told a rich guy to do in the nineteenth chapter of *Matthew*, but which the rich guy from long ago had not been willing to do (as reported by the religious legends).

Eventually, this led to multiple key junctures. One of them was that for a little while approximately April 17th, 2003 I felt that my options to steer out of psychiatric trouble had narrowed down to two: Fundamentalist Christianity or Some Form of Paul Churchland-approved Anti-Religiosity. The "Paul The Apostle versus Paul Churchland" dynamic had reached a breaking point. Suddenly, in the Harris County Psychiatric Center facility as an in-patient, I noticed in the distance a woman singing and speaking in tongues. After a little while, she and I spoke with each other, and it became clear that she had virtually no concern about pressures from the psychiatric authorities against the belief in mystical religious experiences. She would speak openly with me and with others in the facility about her perception that she could hear God speaking to her and that she could interact directly with sharing conversations with God and various angels by thinking thoughts toward them and listening to the responses. Needless to say, the psychiatrists were worried about her influence on each person with whom she spoke. Nevertheless, I chose for a while there to let go of all other religious and philosophical ideas and to wholeheartedly believe in Fundamentalist Christianity. However, after a while, that female patient left, having been either discharged or transferred, and there soon arrived a male patient who was also very assertive, though in a different direction. He openly rejected Christianity, expressing a strong belief in a reality that involves Buddhism, Judaism, and Taoism as being fundamentally and totally real, though it seemed from his expressions that he was a Gentile. (He did not call it "Noahidism," though.) After a while, I was discharged from that facility. In this book, I am declining to state what I clearly remember about what those two individuals told me about their names, and I would encourage personal acquaintances of mine who would happen to know their names to reserve the revelation of their names to be on a need-to-know basis. A huge part of this is that I have no idea at the time of this book's publisher going to press with it whether either or both of those individuals might still be alive, and any major amount of a public revelation of their names would be tantamount to a HIPAA violation (i.e., that would violate their health privacy rights).

Side note: Yes, in putting together this book I am waiving much of what would otherwise be huge portions of my HIPAA (i.e., Health Insurance Portability and Accountability Act of 1996) rights to privacy of my health records, yet I am doing this as part of bearing true witness that could prove helpful to the future of reality. Part of the inspiration was how, a few years ago, one of the most shocking things to appear on TV was that Paris Hilton suddenly revealed to the public that she had years earlier been treated for mental health issues and encountered horrific living conditions in some healthcare facility. For her to have displayed the courage to reveal that, after the previous weird things about how large numbers of media outlets had presented her in mixtures of favorable and unfavorable lights, caused me to rethink many of the assumptions I previously had about what I had experienced as a patient in the psychiatric industry. That was yet another piece of the puzzle of how I overcame years of psychiatric difficulty to become a former psychiatric patient rather than an ongoing psychiatric patient. Of course, as with anyone who is not currently a psychiatric patient, the status of being a nonpatient of that industry could change in the blink of the eye with any societally-deemed misstep if insufficient means of self-defense against treatment were to take place. Also, I believe that some significant percentage of psychiatric patients really should be visiting with psychiatrists and receiving treatments, though there is probably a huge gap between what many of them are receiving as treatments and what would be most ideal for them and for humanity as a whole. I do not claim to know with certainty exactly what changes would best be made, yet I am confident that one of the many roles that this book can play is to provide ample testimony and other tools with which to further people's capacities to improve the psychiatric industry in the very long run.

Back to the performance art idea from the early 21st century, and its transformations. Out of the hospital, I soon embraced an extreme Zen anti-fixed-logic-states orientation toward all idea structures, letting go of many energetic levels of esotericism, Christianity, and everything else, resulting in an exteme equalization of competing ideas and an extreme mellowing out.

However, I remembered having visited True Buddha Temple in Houston briefly on at least two occasions in 2002 (once approximately in the August-to-October period as a casual visit outside of regular group religious practice sessions, and once on the last Sunday of the 2002 NFL regular season for a regular group religious practice session, without having joined yet). Eventually, I returned to them and officially joined in October 2003.

Flashforward to the last few months of 2004. I performed a job search with the help of the Between Jobs Ministry (BJM) program of Northwest Bible Church (NWBC). I had already been involved with BJM during some portions of 2003, yet I finally was on the verge of obtaining a clear career development breakthrough with their help. While working for Accountemps, I networked by BJM arrangement with a man named Clay Collum and a man named Joel Goza. With their help, in late 2004 I obtained an internship offer from KPMG LLP.

I was very excited, believing that this could be the beginning of huge improvements in my life. My father, however, considered what he knew and perceived of the situation and suddenly issued a surprising pronouncement: He advised me to turn down the KPMG offer, continue working with Accountemps, and wait for an offer that could be a better long-term match for my career. At that time, he believed that my personal and professional development had not reached a level that would be conducive for success with the internship converting into long-term career success in general on a reasonably smooth path. Rather, declining the internship, waiting for a better match, then proceeding was what he believed best.

However, when I considered the situation, I decided to go ahead and do the internship and see where things would go. Although I worked hard to get the internship to work well, much of my father's perception of the situation turned out to be true, and KPMG did not offer me a full-time position to start after the conclusion of the internship.

Here are two of the many notable things about that internship. First, there was an unusual dynamic in which a coworker named Lyle Boudreaux had mutliple difficulties with his role there and suddenly disappeared. According to a higher-up on February 22nd, 2005, Lyle Boudreaux had been fired by KPMG LLP on February 21st, 2005 for falling asleep on the job while he was assigned to wait in a room to meet a manager. Second, I had many interesting conversations with the Joel Goza who helped arrange the internship for me; he worked in a different department of KPMG, and he later became the Joel Edward Goza[3] who wrote *America's Unholy Ghosts: The Racist Roots of Our Faith and Politics*. Yes, he is also the person of that name to whom I referred early in Section III of Intermission One as having shared an intriguing conversation long ago.

Flashback now to when I joined True Buddha School. For a while I had explored many different religious practices and emphasized from mid-2000 to sometime March 2003 a mixture of the 1945 Third American Edition of *Tertium Organum* (involving Bessaraboff and Bragdon's translation of that work, with subtle changes from the 1920 Second English edition) and James Legge's *Tao Te Ching* translation. Although my father first introduced me to the aforementioned 1945 edition translation and the aforementioned Ninteteenth-Century translation years earlier than the year 2000, I had difficulties with reading, understanding, and comprehending much from them, reading a few excerpts of the 1945 item circa 1990 and setting it aside and reading a few excerpts of the nineteenth century translation in the mid-1990s and setting it aside. After the strange events of April 12-17, 2000, I soon felt ready to study them. Soon thereafter, I read the *Tao Te Ching* from beginning to end with reasonable comprehension (though it could be debatable whether it is humanly possible for anyone to read that work with absolute clarity, due to its extreme emphasis on pointing toward everything ineffable). After finishing that reading, I completed a reading of *Tertium Organum* (Third American Edition) from beginning to end with reasonable comprehension (though it could also be debatable whether it is humanly possible for anyone to read that with absolute clarity, given its extreme emphasis on all levels of the absolute and all levels of all relative realities).

Returning now to the main flow of supplemental chronologies: The idea of the performance art plan with leaving behind a) a book that would be ostensibly anonymous yet under the surface showing strong evidence of my heavy involvement, b) a disappearance will and testament, c) instructions to authorize various institutions to print additional copies of the book, and d) joining The French Foreign Legion morphed several times in portions of mid-2002 to March 2003. In March 2003 things started to become very weird, emanating from U.S. activities in Iraq and trillions of other things. Some of my experiences from March 2003 to May

2005 paralleled very closely what would become the *All Things Under and Over the Sun and Stars: An Enigma in Twenty-Three Stages* Chapter Twenty description and the *All Things under and over the Sun and Stars: Enigmas in Various Stages* Chapter Sixteen description of the odyssey of personal experiences for the character Heroditus Sterling Blackwell.

Part of that, too, was that I decided in mid-2003 to let go of having much emphasis on *Tertium Organum* and *Tao Te Ching* perspectives, as well as letting go of much emphasis at all on Abrahamic religions[4] of any kind and the vast majority of religious and scientific perspectives in general, as part of wholeheartedly embracing the religious, scientific, and philosophical perspectives, techniques, and technologies of Buddhism. During some of those times, I came up with a main plan to build up wealth and capabilities, then write a somewhat-fully anonymous magnum opus in 2010 and walk away from much of my previous life while retaining most of my financial resources and identity, moving to some foreign nation or another, possibly India.

However, during the KPMG internship, my plans for what to do about the attempt at making a powerful anonymous book and coordinating it with other things as bizarre performance art changed several times to other variations of possible plans. Eventually, between many factors and much of my own initiative, I settled in May 2005 on a plan to put together a semianonymous book to be released in mid-2005 on an extremely limited basis and to be the official publisher of that work. Yes, at some regular ways that some people could conceive of how to coordinate realities, it makes sense to say that I am the author of *All Things Under and Over the Sun and Stars: An Enigma in Twenty-Three Stages* (the writings of which involved the entirety of the researches and developments of my life up to the June 3rd, 2005 finalization of the first edition's first printing's text, and which became published on June 4th, 2005). However, I believe that in many ways to hold onto that idea much at all would ignore the degrees to which it involved channeling through attempts at telepathic mind-melding séances to cowrite it with multitudes of dead authors, who, for all we know, might be capable of acting as very literal ghost writers. Therefore, yes, it makes much sense at some esoteric ways that people could conceive of how to coordinate realities to say that I am not the author of *All Things Under and Over the Sun and Stars: An Enigma in Twenty-Three Stages*, I am merely one of the individuals among the mutlitudes of beings who collaborated on the creation of that work. That being said, I was the person who did all the typing on the keyboard and all the hitting of the save button as parts of creating it.

Although in the regular sense no one has created a second edition of that book up to the time of the publication of the first edition of this book (i.e., *Science, Religion, Politics, and Cards*, with its first publication ocurring in the third quarter of 2023), in some alternative sense *All Things under and over the Sun and Stars: Enigmas in Various Stages* (first published very early in the first quarter of 2023) serves as a book clearly authored by me and carrying with it a highly revised version of the 2005 book as an alternative to there ever having to be a second edition of that 2005 book. There could be incrcdibly many alternatives to thse ways of conceiving things, though.

My friend Dan Mailman, whom I first met through True Buddha Temple in Houston, received one of the printed copies of the second printing of the first edition of *All Things Under and Over the Sun and Stars: An Enigma in Twenty-Three Stages*. I handed that copy to him in early 2006, away from True Buddha School facilities, to the best of my memory, specifically, at some coffee shop in Houston, TX. The 2005 printings were of two types: 1) printing to paper, then using lattices of staples to generate binding the book together sans cover and 2) saving multiple word processor computer files to unified presence (for example, via CD-ROM). The 2005 first printing is a book with no cover, or alternatively conceived, it is a book in which the first normal printed page is both an 8.5"X11" basic paper page and the front cover of the book; that page, by the way, is the copyright notice for it. Although I am the copyright claimant in the 2005 registration of it with the U.S. Copyright office, and although Shunyata appears as the version of Anonymous on the by-line, the copyright line on the first page of that book shows the copyright to be by Universes including Russia and America. The early 2006 second printing had several subtle and several not-so-subtle changes from the 2005 first printing, yet the story text remained the same, with the exception that some electromagnetic anomaly caused a few words in Chapter One Revisited to have a huge font.

Dan Mailman did not seem particularly impressed by the book, as he mentioned to me on some occasions subsequently that he had not read much of it at all after receiving it, and that, as of something like 2007 he had placed it in a box hidden away somewhere in a house in Kerrville, TX.

In contrast with that, I gave one of the CD-ROM copies of the 2005 first printing of it in approximately late July or early August 2005 to Tienling Chen, who a while later, sometime circa 2007, said to me, either verbatim or nearly so, "James! Though Einstein discovered relativity, you discovered absolutivity!!"

3. After I watched significant portions of The Second Impeachment Trial of President Donald John Trump live on TV in February 2021, I found myself somewhat bewildered about some of the alternate ways of considering what in the world had just happened to our country. One of the things I did in the immediate wake of that was to purchase eBooks of Joel Edward Goza's *America's Unholy Ghosts: The Racist Roots of Our Faith and Politics* (though with major skepticism toward many of its premises) (see p. 269), Mark R. Levin's *Liberty and Tyranny: A Conservative Manifesto* (as a set of counter-premises to many of Goza's premises, as well as a supplement to how a few years earlier I had checked out *Ameritopia* from a library and read it from cover to cover, setting out to study it with a few grains of salt, though fewer grains of salt than with the aforementioned Goza book), and Rev. Fr. Paul O'Sullivan's *The Holy Ghost, Our Greatest Friend: He Who Loves Us Best*. Although I have only studied fractions of those three books in the over-two-years of time from when purchasing them until completing the composition of the first printing format of this book, I believe that comparing and contrasting early portions of each of them with early portions of each of the other two, and, to various degrees, with everyone and everything else, has proven fruitful in a good way.

4. Grand Master Sheng-yen Lu himself, via portions of *Dharma Talks by a Living Buddha* (as authored by him and translated by Janny Chow), brought back a small part of Abrahamic religion(s) layered together with some Buddhist Eightold-Noble-Path sometimes-rigid emphases to me after the mid-2003 shift to extreme Zen anti-possession-of-any-rigidly-fixed-logic-structures orientations. His strong emphases at times on the value of how people can, over extended portions of space and time and mind, commit to absolute bearing

of true witness by speech and writing or else remaining silent, brought back a super-high-strength presence of an idea structure tantamount to what some might consider one half of one verse (e.g., the first half of *Exodus* 20:16) and others might consider one alternate full verse adjacent to the entirety of many traditions: "Thou shalt not bear false witness." That was a big part of why from sometime circa late 2003 to circa early August 2008 I practiced an absolute extinction of any intentional lies of commission of any kind, including the total extinction of even the socially condoned so-called white lies in addition to the socially taboo black magic black lies. Therefore, yes, I had completely avoided the full range of intentional lies of commission during that period, from some forms of black magic to some forms of white magic and from black lies to white lies. Some people such as Paul Churchland would at some prior times such as the early 1980s contend in some portions of writing that all forms of black magic and all forms of white magic, as well as the entirety of what many might consider religion and pop psychology, were rooted in lies of considering normal human ideas of reality, perhaps even ranging from Early Empiricism to Critical Idealism, as being more fundamental to reality than Neuroscience-Combined-With-Materialism as a set of ideas of reality. However, others, such as Padmasambhava and Paul the Apostle, have presented expressions, as passed down through the ages, that there can be black truths and white truths in much of what people might otherwise call black magic and white magic, including what some would later consider the normal human ideas of reality, perhaps even ranging from Early Empiricism to Critical Idealism.[5]

In portions of April-to-May 2005, I brought back higher energy states for various manifestations of Christianity, Esotericism, and other things, as part of the transitions. The disappearing, reappearing, and other dynamics started to become very active in May 2005 with respect to how I would relate to Noahidism, Christianity, Taoism, and many other sources of orthodoxy, orthopraxy, heterodoxy, and everything else.

When I did decide to bring back the capacity for me to use intentional lies of commission about 7/12 of the way into the year 2008, after a hiatus of several years, it was with the understanding to reserve it for select extremities; for clarification, here is an excerpt of something I wrote to Jennifer Dai and other Facebook friends at about 11:26 AM U.S. Central Daylight Time on June 14th, 2023, in response to her wishing me a happy 47th birthday: "From around late 2003 to about late July or early August 2008 I acted out a several-years-long total extinction event of intentional lies of commission (i.e., extinction of intentionally stating anything deceptive at its core vis-a-vis basic regular reality) as part of what I believed best in terms of training with True Buddha School methods. Then, I brought back on an exceedingly extremely rare basis, the capacity for intentional lies of commission, but only if I would at some core level know it to be a case of extreme warfare between good and evil with the use of it contributing to the good or if it would have a similarly clearly ethical basis."

Meanwhile, in keeping with much of Eastern Mysticism and much of Western Mysticism, I have believed for many years that lies of omission (i.e., presenting to others a major risk of them missing something of huge importance by intentionally withholding from them vital information) are among the things in some ways unavoidable for beings to do and, in other ways, even when avoidable, among the things completely ethical and fair game for beings to do.

For example, consider if someone were to take a

known drug dealer on a trip to a region where getting caught dealing drugs illegally could lead to authorities legally imposing the death penalty, such as 1994 Hong Kong. The someone could then warn the drug dealer that doing such dealing could have harmful consequences and that said dealer should limit the dispensing of drugs to ethical ways of such distributions, *yet choose to withhold* from *speaking aloud* to the known drug dealer that the legal system in that region is known to execute people who wind up being convicted of drug trafficking.

In that case, many rational and reasonable observers would consider it a lie of omission to not bother to mention that key detail aloud. However, the pusher would have ample opportunity on his or her own to look into what kind of legal situation he or she would be venturing into while visiting such a jurisdiction.

In addition to the fact that ignorance can be lethal, as many have expressed over the centuries, ignorance is no excuse before the law.

There are degrees to which the preceding narratives as perveived by a given beholder may have many half-truths, three-quarter-truths, almost-total-truths, and total-truths circulating through them, and that is one of the vital challenges involving the difference between fiction and nonfiction: Fiction, by virtue of the fact that it calls itself fiction when presented as such, helps its readers to not hold on too tightly to the idea structures that it presents, at the risk of encouraging its readers perhaps a little too much to not take it sufficiently seriously. Nonfiction, in contrast, by virtue of the fact that it typically calls itself nonfiction, can at times give its readers an extra risk of holding on too tightly to the idea structures it presents, though it exhibits a strong tendency to encourage its readers to take it sufficiently seriously.

Scientific and religious literature in the long run tend to present vast arrays of idea structures that can often go to total war with one another, placing political leaders and everyone else in many precarious situations, yet they also afford many opportunities to find and utilize solutions to problems.

5. Consider how it is that many people could strongly agree, strongly disagree, or find themselves somewhere between regarding the page 276 sentence for upon which this footnote makes commentary.

Next, consider the degree(s) of reality-versus-unreality that your mind(s) and the minds of others might decide to attribute to the following alternative supplemental statements:

- Thou art manifestations of black truths and white truths, both of which sometimes exhibit black magic and white magic, including normal and supranormal ideas of reality.

- Various beings have expressed that there can be both white truths and black truths in white magic, black magic, and the rest of the full range of grayscale magic, including what many might consider normal, abnormal, and paranormal realities.

- Though many have fought each other over defining, alternatively defining, and undefining, consider both your motives and others' motives for applying the force of consciousness to choosing and utilizing various definitions and undefinitions of words, people, places, things, ideas, and beyond.

INDEX

Freemasonry 95-96
        --see also Scottish Rites Freemasonry

Game of Death: See project on which initial filming began in
        1972 and Bruce Lee, Dan Inosanto, Kareem Adbul
        Jabbar, and others collaborated... yet which did not
        reach movie theaters until over half a decade later
Garry (ambiguated) 152
Gary (ambiguated) 152
Genesis 138
Gentiles 51
Gerry (ambiguated) 152
Gerrymander (ambiguated) 152
Gervaise (ambiguated) 152
Good(s) (ambiguated) 152
Gore, Al 248
Goza, Joel Edward 100-101, 269-270, 275

Halahala 133-135
Hammer of God, The 247
Henry VIII, King 191
Hevajra 141
Hilton, Paris 268
Hinduism 90
        --see also, Hindus
        --see also, what some consider its opposite, one of
        its offshoots, or something else (or some hybrid of
        two or more of the aforementioned categories of
        items, or beyond all of that), Buddhism
Hindus 51
        --see also, Hinduism
HIPAA (also known as The Health Insurance Portability and
        Accountability Act of 1996) 268
Howard, Queen Katherine 191
Hoppe, Carmen 95
Hoppe, Maurice 95
hydrogen bomb(s): see thermonuclear weapons
Hydrogen-2: see Deuterium
Hydrogen-3: see Tritium